T0182119

Johnstown's Flood of 1889

Neil M. Coleman

Johnstown's Flood of 1889

Power Over Truth and The Science Behind
the Disaster

 Springer

Neil M. Coleman
Department of Energy and Earth Resources
University of Pittsburgh at Johnstown
Johnstown, Pennsylvania, USA

ISBN 978-3-030-06995-7   ISBN 978-3-319-95216-1   (eBook)
https://doi.org/10.1007/978-3-319-95216-1

Printed on acid-free paper

This Springer imprint is published by the registered company Springer International Publishing AG
part of Springer Nature.
The registered company address is: Gewerbestrasse 11, 6330 Cham, Switzerland

For Mara, Kaija, Erika, and Fiona

Acknowledgments

I wish to express sincere appreciation to Ms. Kaytlin Sumner, Archivist of the Johnstown Area Heritage Association, for helping me access the wealth of archival material they collect and preserve for the history of Johnstown. Ms. Marcia Kelly assisted me with images of the viaduct trestle and the South Fork Dam breach. JAHA also maintains the 1889 Flood Museum, a short walk from the Amtrak station in Johnstown, and the Frank & Sylvia Pasquerilla Heritage Discovery Center on Broad Street. I also appreciate access to the microfilm records about the South Fork and Hollidaysburg dams maintained by the Pennsylvania State Archive. They have many records from the years of the Canal Commissioners, including the field notebooks of the supervising engineers who built the South Fork Dam and the Eastern Dam near Hollidaysburg. I am most grateful to Musser Engineering, Inc. of Central City, Pennsylvania, and Professor Brian Houston (University of Pittsburgh at Johnstown) for their GPS analyses of key elevations at the South Fork Dam. The analysis of dam remnant elevations and former lake levels would not have been possible without their contributions.

Thanks to Mr. Timothy H. Horning, Public Services Archivist at the University of Pennsylvania. He provided valuable help in locating documents and images related to the John Parke family history and Parke's classmates, as chronicled in Penn's publications and archival files. Thanks also to Mr. Richard Gregory, author of "The Bosses Club," for sharing some details about the Eastern Dam and for discussions about the floods of Johnstown.

I am grateful to Ms. Meike Gourley, Museum Assistant with the Historical Society in Beverly, Massachusetts, who helped research details about Leroy Temple. I greatly appreciate the help of Ms. Laurel Racine, Chief of Cultural Resources, who provided access to archival material about the Francis family in Lowell and for general information about the canals and tours provided by the National Park Service. Thanks also to the staff and management of the Johnstown Flood National Memorial for their educational and preservation efforts, and for their permission and assistance to do research in hydrology and geomorphology at the historic site of the South Fork Dam. Ms. Nancy Smith of the National Park Service located and provided historic photographs of the South Fork Dam from the time the NPS first acquired the property.

Very special thanks to my friends and colleagues, Stephanie Wojno and Nina Kaktins, for permitting the use in this book of material from our 2016 paper in the journal *Heliyon*. Ms. Wojno contributed numerous figures for this book and co-authored and greatly enriched Chapter 13 about "The Forgotten Dam" near Hollidaysburg.

Recommended Reading

Carnegie A (1920) Autobiography of Andrew Carnegie (with illustrations). Houghton Mifflin Co., New York and Boston, 385 p

McCullough DG (1968) The Johnstown flood. Simon & Schuster, New York, 302 p

Qing D[1] (1998) "The River Dragon has Come! The Three Gorges Dam and the Fate of China's Yangtze River and its People." Thibodeau JG, Williams PB (eds) [*see footnote*]

Sharpe EM (2004) In the shadow of the dam – the aftermath of the mill river flood of 1874. Free Press, New York 284 p

Website of the Johnstown Area Heritage Association and the Johnstown Flood Museum; at http://www.jaha.org/attractions/johnstown-flood-museum/flood-history/

Website of the Johnstown Flood National Memorial, National Park Service; at https://www.nps.gov/jofl/index.htm

[1] Dai Qing is the daughter of a revolutionary martyr. As a journalist she was imprisoned for her writings about the Three Gorges Dam. She is now forbidden to publish in China but continues to advocate for freedom of the press, government accountability, and dam safety. I highly recommend her works.

Prologue

The literature of the Johnstown flood consists of several films and hundreds of articles and books, most of which focus on the human tragedy, the horrors of the flood wave descending on Johnstown and its neighboring boroughs, scenes of mass destruction, and the heroism and suffering in the aftermath. The public will always be fascinated by the morbid and the macabre. Yet strangely, almost no quantitative science has been done on the 1889 flood, to search for the true cause of the dam breach that spawned the disaster. Perhaps there is a psychology of tragedy that causes professionals to turn away from analyzing terrible events in detail. But such analysis can save lives in the future, such as the engineering studies of the speed of collapse of the World Trade Center towers after exposure to impact and intense fire.

The study of dam safety is crucial to our society. There are tens of thousands of dams in this country, some of which have been labeled high-risk dams. If they were to fail, the lives of hundreds, thousands, and even tens of thousands would be at risk. Sadly, the lessons of the Johnstown flood and other dam breach disasters have yet to be learned. In 1975 in China, runoff from torrential rains breached the Banqiao and Shimantan dams, causing the domino collapse of scores of smaller dams and levees below them. Hundreds of thousands perished in the flooding and its aftermath. Such tragedies have only one redeeming value – they serve as examples of events to anticipate and avoid – what to do and what not to do. They teach important safety lessons to individuals and society as a whole.

The American Society of Civil Engineers (hereafter the ASCE) launched an investigation immediately after the 1889 flood. Three members of the appointed committee were among the finest hydraulic engineers in the USA. The breached dam was owned by the elite South Fork Fishing and Hunting Club of Pittsburgh. Some of its members were among the wealthiest industrialists in the nation. They had the power to influence the investigation and even the inner workings of the ASCE. It is unfortunate that even when public safety is at risk, power can overrule science and engineering. The fate of the ASCE investigation report is a tale of power over science.

Contents

About the Author

Neil M. Coleman served as a Senior Staff Scientist for two federal advisory committees, the Advisory Committee on Nuclear Waste and the Advisory Committee on Reactor Safeguards. He retired in 2010 from the Nuclear Regulatory Commission in Rockville, Maryland. A Navy veteran, he served on the carrier USS Constellation, later earning the M.S. degree in Geology from the University of South Florida. Mr. Coleman is a Professional Geologist in Pennsylvania and currently teaches geophysics at the University of Pittsburgh at Johnstown. He has numerous peer-reviewed publications, including lead authorship (with Victor Baker) of a chapter in the 2009 book "Megaflooding on Earth and Mars." His continuing interests include the safety of the nation's dams, geophysics, the geology and paleo-hydrology of Mars, and research with colleagues on the watershed of the Little Conemaugh River.

List of Abbreviations

ASCE   American Society of Civil Engineers
Club   South Fork Fishing & Hunting Club
DCNR   PA Dept. of Conservation and Natural Resources
DEM   digital elevation model
ENARJ   Engineering News and American Railway Journal
Eng. News   *Engineering News*
ESWP   Engineers Society of Western Pennsylvania
EXPO   Exposition Universelle in Paris in 1889
GSA   Geological Society of America
JAHA   Johnstown Area Heritage Association
LiDAR   portmanteau word combining "light" and "radar"
NAVD   North American Vertical Datum
NPS   National Park Service
PA   Pennsylvania
PASDA   Pennsylvania Spatial Data Access
PRR   Pennsylvania Railroad Company
AO   former PRR telegraph tower about 1.5 miles east of East Conemaugh
Tribune   *Johnstown Daily Tribune*
USDA   US Department of Agriculture
USGS   US Geological Survey

Chapter 1
Introduction

On the last Friday of May, 1889, the South Fork dam was overwhelmed by floodwaters and breached. The collapse of the dam drove down the valley of the Little Conemaugh a massive flood wave that destroyed Johnstown and its neighboring boroughs, killing more than 2200 people. Images of the destruction were broadcast in newspapers around the country and in Europe, bringing into peoples' homes a vivid sense of the terror the victims felt as the wave approached, just as Mathew Brady's photos of Civil War battlefields brought home the true carnage of war in ways that lists of the dead and wounded could never convey. Early reports that 10,000 or more people had perished were thankfully overstated, but the actual losses were wrenching and beyond belief because this was not war.

The newspapers and engineering periodicals harshly criticized the source of the flood, a dam and lake owned by the elite and aloof South Fork Fishing and Hunting Club. They said the Club had done a shoddy repair on the dam that led to the dam breach flood. In fact none of the Club members or its workers who repaired the dam were ever held legally liable for the deaths and property destruction.

As soon as rail lines were running again with temporary repairs, the American Society of Civil Engineers (hereafter abbreviated as "ASCE") dispatched four of the finest engineers in the United States to inspect the remains of the dam and investigate the cause of its demise. Three of them were highly respected hydraulic engineers, including the renowned James B. Francis, who some consider to be the "father" of hydraulic research in this country. The other hydraulic engineers were the renowned William Worthen and his gifted protégé, Alphonse Fteley. The fourth engineer, Max Becker, was a prominent railroad man out of Pittsburgh. The team prepared a report dated January 15, 1890, with detailed calculations, an appendix, and a discussion section. Francis' name headed the list of authors, which made it appear he chaired the Committee. He was now retired and an emeritus member of ASCE. The release of this report was highly anticipated, but it was immediately suppressed, not given to the press, the public, or even so far as I can determine to other members of ASCE. It finally was released more than 2 years after the disaster, after most court cases had ended and much of Johnstown was rebuilt. Despite being

© Springer International Publishing AG, part of Springer Nature 2019
N. M. Coleman, *Johnstown's Flood of 1889*,
https://doi.org/10.1007/978-3-319-95216-1_1

state-of-the-art for that time, the report bears the signs of being watered down and sanitized. And why the long delay, when the findings could pertain to the safety of other dams where the men, women, and children below them were at risk? Just a few weeks after the flood, Fteley, who was a vice president of ASCE, had called for prompt publication for that reason.

Even today few people know that the Club was officially exonerated by the ASCE report. The report presented calculations that showed the dam would have been overtopped and destroyed in 1889 even had it been repaired as originally built in the 1850's. Their report and its findings were not formally challenged in the scientific literature until 2016, when our team of geologists analyzed the hydraulics of this infamous dam breach. Science now reveals the true cause of the Johnstown flood and examines the history of that 1891 report and the engineers who wrote it. The hydraulic engineers were truly exceptional, even brilliant, so their controversial findings bear scrutiny in our modern age when dam safety remains a priority for the protection of our people.

Over 5 years a team of geologists sifted through the historic and scientific evidence and analyzed the flood of 1889. Professor Emeritus Uldis Kaktins, Stephanie Wojno, and I found clear hydraulic evidence that the dam would have survived the 1889 storm had it been properly repaired. Even though the city was and still is flood prone, the destruction of Johnstown should never have happened in 1889. We critically examined the ASCE report that acquitted the Club of blame. It contains discrepancies and lapses in key observations, and relied on excessive reservoir inflow estimates. They expressed confidence that the dam failure was inevitable. But that conclusion was inconsistent with information gathered by the Committee themselves in June, 1889. The report also shows evidence of meddling by people from outside the Committee. What truly happened to this important investigation?

Interwoven in this narrative are remembered conversations with the late Uldis Kaktins, one of my colleagues in this research, who was brought to this country as a young child from war-torn Europe. He became an Army officer and war veteran, eventually a respected and cherished emeritus professor of geology. His love of geology, deep insights, and gentle humor touched the lives of thousands of students. Kaktins studied the floods of Johnstown throughout his career and inspired this work and other research in hydrology. He passed away in early July, 2016, a few weeks after our dam-breach paper on the flood was published in the journal *Heliyon*. This new book would not exist without the backdrop of his years of research on the floods of Johnstown, and his dedication to the passionate teaching of geology and hydrology. My career was inspired by this gentleman of science. As Isaac Newton observed in a centuries old letter to Robert Hooke, "If I have seen further it is by standing on the shoulders of Giants."

I have gone beyond our earlier work to study the background of the engineers who investigated the flood and who wrote the ASCE report. Biographical sketches of these men, their accomplishments, and their families are included here. What was going on in their lives at the time of the flood, and who were they associated with? What may have happened in 1890 and 1891 to delay the report *for 2 years*, and who in or out of the ASCE had the motives, the power, and the guilt to intervene and influence its release and conclusions? Who wanted to control the judgment of history?

It is sometimes necessary for scientists to delve into history, to unearth the past, especially when the real nature of things was concealed in subtle ways. But nineteenth century clues were left as they always are, and a trail of crumbs led to the true story of the cause of the Johnstown flood and the engineering study that tried to clear the "Boss" Club of blame. In the end, powerful men found a way to control the release and probably also the content of the investigation report. But first we will start at the beginning, the world as it was in 1889, the gathering storm clouds, and the history of the South Fork Dam.

1.1　The U.S. and the World in 1889

The United States and the world were changing rapidly in 1889. On February 22, President Cleveland signed the bill to admit Montana, Washington State, and the Dakotas to the Union; they officially became states later that year. In March, Ferdinand von Zeppelin patented the hydrogen-filled airships that would bear his name and in WWI would terrorize the skies over England. Alexander III reigned jointly as the Emperor of Russia, the King of Poland, and the Grand Duke of Finland. The first golf course opened in the U.S. Charlie Chaplin was born on April 16th, and 4 days later so was Adolph Hitler. The race across the prairie, the Oklahoma land rush, began on April 22, five and a half weeks before the Johnstown flood.

A new dance, the two-step, gained rapid appeal as John Philip Sousa debuted his "Washington Post March" on June 15th, and dancers realized it was a bold accompaniment to their new moves. Also in June, Vincent van Gogh painted "Starry Night," one of 142 works during his year at the asylum in Saint-Rémy. Later that year he painted the somber "Wheat Field in Rain." On June 3rd the Canadian Pacific Railway was completed from coast to coast, an artery of steel rails that would mean as much to Canada's economic future as the first Transcontinental Railroad in 1869 meant to the U.S. A great fire destroyed 25 blocks in downtown Seattle on June 6th, and cable car service began in Los Angeles 2 days later. The Johnson County War in Wyoming was instigated that summer with the rare hanging of a female rancher, Ella Watson, and her husband.

On July 1st Frederick Douglass was named the U.S. Minister to Haiti. Also in July, John L. Sullivan knocked out Jake Kilrain in an epic, 75 round bare-knuckle fight. In Brazil the Emperor Dom Pedro II was deposed in the Brazilian Revolution of 1889. Astronomer Edwin Hubble was born on November 20th. His amazing discoveries would reveal the true scale of the universe. And on December 4th, explorer H. M. Stanley emerged from the African jungles to reach Bagamoyo on the Indian Ocean.

In the United States the 1880's mostly represent positive and uplifting change, with electrification coming to the fore and the era of small-scale water power declining. No longer would industries have to be sited by streams and rivers to meet their power needs. It was the start of a golden age in engineering and recognition of the importance of clean water supplies for populations in cities and towns. But there was retrograde movement as well, with rampant discrimination, especially in the southern states, and the imposition of Jim Crow laws in the 1890's and beyond.

Benjamin Harrison was President of the United States on the day of the 1889 flood. Young Henry Ford had been married for a year and was making a living at farming and running a sawmill. Eastman Kodak became the first company to commercially produce a flexible celluloid film, a great advance over the rigid plates then in use. Orville and Wilbur Wright were young men aged 17 and 22. Wilbur had been dreaming of flying machines since childhood. Three months earlier in Dayton, on a press made by Orville, they had published the first edition of their newspaper, *The West Side News*. One of their June papers mentioned the flood in Johnstown (McCullough 2015).

The Eiffel Tower in Paris was newly built on March 31st, 1889. It was the centerpiece of the Exposition Universelle that was about to begin. The Tower opened to the public on May 6th, the day of the grand start of the EXPO which ran through October. Annie Oakley joined Buffalo Bill's *Wild West Show* for many performances there.

1.2   Carnegie and Pitcairn

Andrew Carnegie and his wife were attending the Paris EXPO when they and the rest of Europe got the news via transatlantic cable of a terrible flood in Johnstown. And he learned it had been caused by the failure of the dam above South Fork, at the summer retreat and private club where he was a member. Yet his autobiography published in 1920 does not even note their trip to the EXPO, which was surely memorable. The book has no mention of the South Fork Fishing and Hunting Club, and includes the word "Johnstown" only twice, both times in relation to the steel industry. The year 1889 appears in his book only three times, twice with regard to a steel worker union agreement, and the third instance the year one of Carnegie's friends died. He wrote about a disaster in his youth, when a "great flood destroyed all telegraph communication between Steubenville and Wheeling" on the Ohio River, but Carnegie's book does not even mention the far more disastrous Johnstown flood, which killed more than 2200 people.

The things left out of a narrative are often the most telling. Carnegie clearly intended to ignore any perceived connection he had to the Club on the lake and the destruction of Johnstown. The guest ledger of the South Fork Fishing and Hunting Club survives to this day, but 73 pages are missing. After the pages were ripped out, the final preserved entries were from June 1886. There is therefore no evidence from the ledger that Carnegie ever visited the Club, although his name persists in the list of members in the back.

Carnegie undoubtedly wanted his legacy to remain clear of any taint, and indeed his record of philanthropy is admirable. Thousands of libraries were built with his support, the first in 1883 in his home town in Scotland. Carnegie's philanthropy continued for the rest of his life and through posthumous gifts. His actions and writings about social responsibility inspired others to donate. But it should be remembered that he and his partners had made a fortune selling steel armor for US Navy ships, and Carnegie had tried to get his partners to retool to go after more military work. The partners balked. But his writings and speeches truly pressed for peace. In 1910 he gifted $10 million toward a "peace trust," which brought his charitable bequests to date up to the extraordinary amount of $200 million (Nasaw 2006). The peace trust became the Carnegie Endowment for International Peace, the kind of benefaction Carnegie hoped could prevent future wars. He truly believed that well-directed philanthropy by the wealthy could achieve this.

But sadly for Carnegie, and especially for the millions of people who would be affected, World War I was just around the corner. Philanthropy could not stop it. He became deeply depressed after that terrible war began. As his wife wrote in the preface to his autobiography, he stopped writing his journals and no longer felt able to do the things he loved best; fishing, golfing, and swimming. After his death in August of 1919, his widow published his autobiography for all to read. Despite Carnegie's exceptional legacy of philanthropy, he will always remain partly in the shadow of the Johnstown Flood.

Robert Pitcairn was Carnegie's boyhood and life-long friend. Both were youthful immigrants from Scotland. They met in Pittsburgh and both worked as messengers at O'Rielly's Telegraph Company for the salary of 2½ dollars per week. Ever after, to each other, they were "Bob" and "Andy" (Carnegie 1920). By 1889 Carnegie had been married for 2 years and was mostly retired, traveling extensively and continuing his extraordinary philanthropy. Pitcairn was now the powerful Superintendent of the Western Division of the Pennsylvania Railroad Company, a corporation so dominant in commerce and politics it was simply called "The Company." Both men were also members of the South Fork Fishing & Hunting Club and the exclusive Duquesne Club of Pittsburgh. Pitcairn was deeply religious, Carnegie was not, but both were sensitive to their reputations, how they were perceived in the world and how they would be viewed by posterity. No doubt they believed their legacies could be haunted by the thousands who perished in the 1889 flood.

Their legacies were indeed imperiled soon after the disaster when coroner juries from Cambria and Westmoreland Counties declared the obvious, death by violence and death by drowning caused by the breach of the South Fork dam. The Cambria jurors went further to say that the owners of the dam were culpable and responsible for the loss of life and property (McCullough 1968). The elite Club, its members mostly from Pittsburgh, had virtually been tried and convicted in newspapers. Professional periodicals like *Engineering News* carried reports from civil engineers who visited the ruined dam and toured the flood path. They harshly criticized the dam repairs supervised by Club founder Benjamin Ruff, especially the failure to involve qualified engineers. And now the ASCE investigation had begun by a committee that included three of the foremost hydraulic engineers in the country. Their conclusions when published could condemn or exonerate the Club and its members.

Carnegie, Pitcairn, and the other powerful members of the South Fork Fishing & Hunting Club were men accustomed to getting whatever they wanted. They had every reason to try to protect their reputations by influencing the timing of release and conclusions of the ASCE report. The three hydraulic and dam engineers, Francis, Worthen, and Fteley, were not railroad or steel men. No direct pressure could be brought against them. But the fourth member of the investigation was Max Becker, who now lived in Pittsburgh and was the Chief Engineer of the Pittsburgh, Cincinnati & St. Louis Railway. He had virtually no experience with dams or hydraulic engineering. But in 1889 he happened to be the President of ASCE and was inserted as chairman of the investigation committee. And by chance his railway was controlled via stock ownership by Robert Pitcairn's company, the powerful Pennsylvania Railroad.

References

Carnegie A (1920) Autobiography of Andrew Carnegie (with illustrations). Houghton Mifflin Co., New York/Boston, p 385
McCullough DG (1968) The Johnstown flood. Simon & Schuster, New York, p 302
McCullough DG (2015) The Wright brothers. Simon & Schuster, New York, p 320
Nasaw D (2006) Andrew Carnegie. The Penguin Press, New York, p 878

Chapter 2
The Gathering Storm – "A Shower of Fishes"

In 1890 the collection and analysis of weather data was transferred from the War Department to the new Weather Bureau, under the Dept. of Agriculture. But in 1889 weather data from around the country were gathered via telegraph by the Signal Corps of the War Department, commanded by Brigadier General A. W. Greely, the Chief Signal Officer of the Army. The Signal Corps plotted the surface tracks of low pressure systems each month.

Unusual weather had already been seen in May. Fish rained from the sky in Kansas, according to an official weather report (Dunwoody 1889a, p 126):

> Wichita, Kansas: during a thunder-storm which occurred the afternoon of the 10th [of May] *a shower of fishes*, from one to four inches long, fell at the Burton Car Works, four miles north of this city. They covered the ground in thousands. One, brought to police headquarters, was a small catfish about three and three-fourths inches long, such as abound in the streams hereabouts.—Report of Signal Service observer.

This astonishing report could only be explained by a small tornado touching down on a nearby stream, swirling water and small fish into the air long enough to drop them nearby. Tornados are indeed more common in the late spring clashes of contrasting air masses. The month of May, 1889 also spawned the first known Atlantic hurricane so early in the year.[1] That storm gathered north of the British Virgin Islands, reached hurricane strength on May 20th, then paralleled the East Coast but did not make landfall. Less than a month later a second hurricane formed, but it weakened into a tropical storm that passed over northern Florida and into the Atlantic.

---

[1] The earliest recorded hurricane in the Atlantic is now thought to have spawned in early January, 1938.

N. M. Coleman, *Johnstown's Flood of 1889*,
https://doi.org/10.1007/978-3-319-95216-1_2

2.1   The Coming Storm

Sandwiched in time between these cyclones, the Signal Corps began tracking the 8th low pressure system of the month on May 28th, first charting its center in south-western Kansas near the Oklahoma panhandle. This system moved fast, passing through Kansas-Oklahoma on the 28th and Missouri-Arkansas and northern Mississippi-Alabama to Tennessee on the 29th. By the 30th the low pressure center was over Kentucky where it split into two identifiable lows. One tracked eastward through West Virginia and Virginia-North Carolina until it reached the Atlantic on the 31st. The other low pressure center moved northeastward across southern Ohio, then on May 30th it drifted northward along the Ohio-Pennsylvania line. Cincinnati recorded a low barometric pressure of 29.58 inches on the morning of May 30th. The lateral pressure gradient was substantial, 30.38 inches being measured at Duluth. The system slowed, creating a nearly stationary weather pattern over east-ern Ohio and western Pennsylvania, then began to turn northwest into southern Michigan where this low further divided into secondary depressions. On May 31st the remnants of one moved west toward Chicago, while the other depression moved northeast over lakes Erie and Ontario toward New England. These movements reflect only the estimated tracking centers; cloud and precipitation bands covered large regions in the northeastern U.S.

Weather system VIII with its secondary depressions was an unusually strong storm that wreaked havoc wherever it went, but especially in the northeast. As is typical of spring storms that begin in the western plains, the track and counterclock-wise rotation of winds brought strong northward flow of warm, moist air off the Gulf of Mexico into contact with cold, dry air over the eastern plains. The circula-tion also drew in large amounts of moisture from the Atlantic Ocean. Large tem-perature differences from northwest to southeast across the Ohio Valley combined with the abundant atmospheric moisture to create perfect breeding grounds for severe thunderstorms and excessive rainfall. Snow fell in parts of Michigan and Illinois. Temperatures over the Great Lakes were as low as 40° and increased to 70° on the Atlantic Coast. Analysts with the Signal Service described the secondary depression moving west over Michigan as "…favorable to the agricultural interests of the Northwest, as it caused a continuation of cloudiness, thereby preventing a destructive frost…" (Dunwoody 1889a, p 108).

In 2 days the storm traversed the zone of highest average tornado risk in the U.S., producing substantial rain. Numerous stations along the path saw rainfalls >2.5″ in 24 hours for May 29 and 30. It must be remembered there were far fewer weather stations in 1889 than in later years. In Kansas, near the storm's origin, six stations had rainfalls >2.5″ (6.4 cm) during May 27–28, with Lebo in eastern Kansas getting 4.6″ on the 28th. Seven Missouri stations noted rainfall >2.5″. Nine inches fell in eastern Missouri at New Frankford. Nine Tennessee weather stations had 2.6 to 3.5," and Louisville got 3″. Five Ohio stations had rainfalls of 2.6 to 6″.

But the most concentrated regional rain fell during the night of May 30–31 on western Pennsylvania and adjacent parts of New York. On May 31st, 6 inches fell at

West Almond, NY. Appendix 1 gives inches of rain for Pennsylvania stations, including the following: Aqueduct, 5.70; Charlesville, 6.71; Coudersport, 5.40; Eagle's Mere, 5.17; Emporium, 5.85; Harrisburg, 6.16; Hollidaysburg, 5.12; Petersburgh, 6.60; Selin's Grove [Selinsgrove], 6.00; and Smethport, 5.50.

The *Monthly Weather Review* (Dunwoody 1889b, p 150) noted that the heavier rainfalls began earlier in southcentral Pennsylvania than in the west and north. For example, the heavier precipitation started at 3 p.m. on May 30th in Bedford, Altoona, and Gettysburg. Farther north and east, in Williamsport, the heavier rain commenced at 9 p.m. It began at 8:30 p.m. in Uniontown, to the southwest, and 10 p.m. to the north in Erie.

On May 30th at the South Fork Fishing & Hunting Club, engineer John Parke reported he had gone outside around 9 p.m. and found the boardwalk wet from a slight rain, but it was not raining at that time. He also wrote that the sky was "... evidently clearing off, but there was still a high wind blowing." (Francis et al. 1891 p 448).

Only two inches of rain had been reported on May 31st from Johnstown itself, but this partial reading was taken before the flood destroyed the station. Mrs. Hattie M. Ogle of Washington Street, age 52, managed the Western Union telegraph office in Johnstown. She had been the Signal Service weather observer since November 1, 1884. Due to high water in the streets Mrs. Ogle and her staff had moved their equipment to the second floor (Shappee 1940). Ogle's final report gave 2.3" of rain by 11 a.m. on May 31st. She was lost in the flood five hours later - her body and that of her 32-year-old daughter Minnie were never identified. The storm total over Johnstown was later estimated at 3 to 3.5" (Blodget 1890).

A key factor for the survivability of the South Fork dam was the time-varying rate of flow into the reservoir. That rate depended on the rainfall intensity and duration and speed of runoff into the streams. Estimates of the time of concentration and time to peak stream discharge are developed in Appendix 3. I estimate a range of time of concentration (t_c) for the South Fork subbasin of 3.6 to 7.3 h, and time to peak discharge (t_p) in the range of 2.5 to 5.1 h. Data for the 1977 flood from the river gage at East Conemaugh[2] suggest that t_p in extreme events upstream covering the South Fork basin would be <7.3 to 8 h.

Rainfalls in parts of the region appear to have reached or exceeded the threshold for 100-year, 24-h storms (USDA 1986 Fig. B-8). Fig. 2.1 shows the approximate rainfall distribution in Pennsylvania during May 30–31, 1889. The areal coverage of stations was quite sparse - the closest weather stations to the South Fork dam included Blue Knob (19 km east) and Hollidaysburg (34 km east). Rainfalls associated with the 1889 event were 9" (1.00 in = 2.54 cm) or less, with the highest measurements being reported at Grampian (8.60"), Blue Knob (7.90"), Charlesville (7.60"), and far to the east at McConnellsburg (8.99"). The station at Blue Knob was just east of the watershed. Heavy rainfalls in Blue Knob and Grampian suggest the influence of a phenomenon known as "upsloping," in which saturated air masses

---

[2]At the time of the 1889 flood the present-day borough of East Conemaugh was the village of Conemaugh.

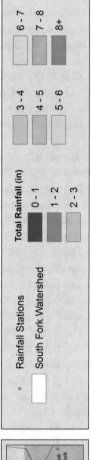

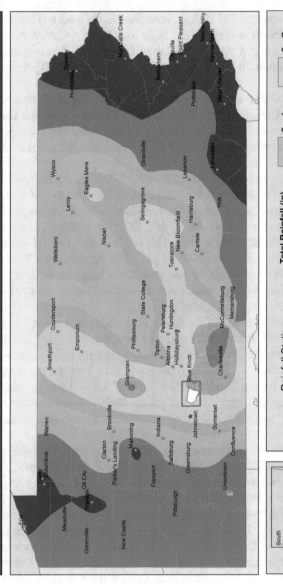

Fig. 2.1 Approximate rainfall distribution in Pennsylvania during May 30–31, 1889 (station data provided in Appx 1). Contour interval = 1.00 inch (2.54 cm) of rainfall. Cities and towns on the map are data locations. Contours developed using a kriging interpolation method with the measured data points. This method is based on the statistical relationship between measured points to produce a predictive surface and provides a measure of accuracy in those predictions (ArcGIS 2011) (sources: Blodget 1890; Russell 1889; Townsend 1890). Credit: S. Wojno

move upward against mountain slopes, begin to cool, and produce large amounts of precipitation.

A plausible average rain total over the South Fork watershed was 6 to 7″, with as much as 7+" in the highest and easternmost parts of the drainage basin. In 1889 there were no precipitation stations in the South Fork watershed, and even today there are few. One anecdotal observation was noted by Francis et al. (1891 p 468). A small dam existed on a tributary to Lake Conemaugh. The dam's proprietor said that "... a pail, which was empty the night before, had 6½ inches of water collected in it from the rain during the night. This was at a point about two miles above the dam." Russell (1889) estimated that about 75% of the total precipitation had fallen by 3 pm on May 31st. Although Francis et al. (1891 p 452–453) believed that the stream discharges into the reservoir kept increasing until the time the dam failed, ".... and no doubt continued to do so for some time longer," the evidence they offer for that assertion does not include local stream observations. They did not document in their report the names of people they interviewed. The main source of stream observations that I found was the Pennsylvania Railroad testimony, ordered by Robert Pitcairn (PRR 1889). Testimonials, mostly from PRR employees who witnessed the flood, were taken in Pittsburgh starting on July 15, 1889. But in support of the ASCE committee, I found no evidence that the PRR gave them those testimonials to review as part of their investigation.

There have been claims that heavy rains had soaked the Conemaugh Valley and South Fork watershed before the major storms on the night of May 30–31. If such storms had occurred they would have saturated the soils in the watershed and promoted more rapid runoff. But *there is direct evidence that the effects of any antecedent storms had greatly diminished.* The Signal Office (Dunwoody 1889a p 118) received measurements from the Johnstown river gage on May 30th and 31st, before it and presumably its observer were swept away by the flood. At 7:44 a.m. on May 30th it read "1.0 foot above low water." That gage reading integrated the effects of any prior storm runoff on the watershed, and suggests that conditions approaching base flow had resumed by the morning of May 30th.

2.2   Cloud Seeding!?

Emeritus Professor Uldis Kaktins had an intriguing and thoughtful theory about the extreme rainfall event in the Johnstown region in 1889. On a rainy afternoon in the fall of 2010 we discussed the precipitation patterns. As rain was beating on the northwest windows of his home, Uldis spread a map out on his dining room table, showing Johnstown and regions to the east and north.

> "Look at this" he said, placing a transparent overlay on the map. The overlay had colored rainfall contours, neatly showing a pattern of enhanced rainfall east of Johnstown.
>
> Uldis tapped a pencil on the map and said, "At one point in the storm the winds had shifted, out of the west. The iron works were in full swing …all those coke-fired furnaces blazing."

I replied, "Yes, but what does the wind have to do with any of that?"

Uldis smiled. "The town had lots of soot in those days. Those chimneys pumped hot clouds of fine ash, particulates, high into the air, condensation points for raindrops. Looks a lot like cloud seeding to me! Inadvertent of course."

He had a good point. The 1889 storm moved into the Johnstown and South Fork region from the southwest, but the winds would have shifted rapidly as the low-pressure region passed. Given a combination of rising air over the Alleghenies with an injected haze of hot particulates from industrial chimneys, the result could readily have seeded the cloud layers and induced greater precipitation.

The iron industry functioned by the large-scale burning of coal and its concentrated energy form of coke. Particulates from these operations were thermally injected into the atmosphere via tall smokestacks. Carbon particulates would have served as ideal nucleation sites for raindrops. Soot output was reduced on the morning of the flood because many iron workers had been allowed to go home due to water running through the city streets. The furnaces would still have been hot, but more importantly, most of the rain had already fallen on the watersheds and therefore any "seeding" had already happened.

More detailed meteorological information from many more stations is available for the buildup to the 1936 and 1977 floods in Johnstown, so the possibility that cloud seeding affected those events could be further investigated. The Seward coal-fired power plant near Johnstown had been operating since 1921 and, along with the then-operating steel mills, would have contributed some atmospheric particulates.

2.3   Regional Storm Damage

The storm systems caused damage over large regions of the northeastern U.S. The June 8th issue of *Eng. News* (1889) gave early reports from Kansas eastward, prepared from dispatches and newspaper accounts. Heavy rains fell over parts of Nebraska and Kansas, wiping out bridges in Leavenworth County. A landslide blocked the tracks of the Santa Fe railroad near Courtney, MO. Heavy stream flows without serious damages were reported from Michigan, but the interaction of contrasting air masses led to late spring snowfall in parts of the state. Snow also fell at Tuscola, IL, over Indiana at Michigan City and Winamac, and east and southeast of Chicago. Farther east in Coburg, Ontario the flooding wiped out some mill dams and bridges. Parts of the District of Columbia and nearby Georgetown flooded, and there was a plan to inspect the foundations of the Washington Monument for possible harm. In Virginia, the large dam of the Petersburg city reservoir breached on June 1st and four bridges washed away. The dam and locks upstream from Fredericksburg were destroyed. Martinsburg, WV experienced one of the worst storms ever seen there. Three bridges between Cumberland and Piedmont on the West Virginia Central Railway were carried away.

As described by *Eng. News* (1889), the greatest ravages were reserved for Pennsylvania, Maryland, and New York State. Railways were severely washed out around Maryland and towns were flooded along the Susquehanna. A newly built bridge over the Potomac River at Point of Rocks was swept out, and at Frederick this river exceeded previous flood heights by 6 to 8 feet. At least 64 bridges were destroyed, including a stone bridge across the Potomac at Shepherdstown. The *Eng. News* (1889 p 528–529) report on Pennsylvania was sobering:

Throughout the State floods occurred. At Lewiston 8 railways and county bridges were destroyed: the water was 9 ft. higher than in 1817. At Lewisburg the railway bridge was carried away and the water and gas works flooded. A highway and wagon bridge, and several small bridges at Milton, were carried away. Newport was flooded with about 7 ft. of water. All the bridges on Tuscarora, Lost, and Cocolamus creeks in Juniata Co. have been carried away, and the Juniata river bridges at Mifflinto[w]n, Mexico, Port Royal and Thom[p]sontown in the same county; the dams, culverts, lock house and waste weirs of the Pennsylvania canal at Mifflinto[w]n were also carried away. Washouts of great extent occurred on the Pennsylvania Railroad, and nearly all the roads in the flooded district suffered more or less from washouts, wrecked bridges, landslides, etc. Traffic was seriously delayed. It is thought that Northumberland Co. and Bedford Co. will each sustain a loss of $50,000 in bridges. Williamsport was flooded with 34 ft. of water in some places, the flood of the Susquehanna being 7 ft. higher than that of 1865. The rainfall is said to have been unprecedented. Harrisburg and Steelton were flooded. The trouble on the Pennsylvania Railroad was the worst ever known, some of the washouts being 300 to 400 ft. long, the tracks flooded and blocked with debris and landslides, and numbers of bridges being swept away. York Co. has lost $100,000 in bridges. At Bradford, Pa., the Allegheny River was higher than it has been for 21 years. In the valley around Mt. Carmel, the mines are being flooded, and the pumps are under water.

In the main parts of Lock Haven and Williamsport, floodwaters were 4 to 8 feet deep. Seventy-eight people reportedly died in the counties bordering the west branch of the Susquehanna River. The loss of so many road and railroad bridges in the northeast disrupted commerce for weeks to come. There was much engineering work to be done quickly to get commerce moving, and now there were new flood levels at former bridge sites for the engineers to ponder in designing better replacements.

Parts of New York State were also hammered by the regional storm. I was dismayed to read newspaper accounts from Corning, NY of the tremendous damage there. Such narratives were mostly lost in the shadow of the Johnstown disaster, and sadly, few remember the stories of lives lost elsewhere. The *Corning Journal* (1889) proclaimed:

The Deluge!

GREAT FRESHET IN CORNING ON SATURDAY, JUNE 1st.

The Water Higher than Ever Before.

AN UNPARALLELED DAMAGE TO PROPERTY.

OTHER TOWNS THROUGHOUT THIS VICINITY FARE EVEN WORSE.

All communication with Johnstown and nearby towns had been lost, with telegraph lines down, until witnesses walked or rode out of the valley on horseback. Early reports suggested that Johnstown had been obliterated and wrongly claimed that 10,000 to 20,000 people (or more) perished in the Conemaugh valley. Writers and editors made such claims in rather cavalier fashion, simply pulling the 1880 census report off their bookshelf. They then estimated growth for Johnstown since

1880 and ginned up a number, assuming most of the citizens had died. But not all the people lived on the floodplain, and the rising water in the streets had fortunately driven thousands to abandon the lowest areas and this saved their lives, greatly reducing the toll of dead and injured. In the immediate aftermath, hundreds of survivors were trapped in the packed, partly floating wreckage above the stone bridge. The longest afternoon and night of their lives had begun. Difficult weeks lay ahead for the survivors, many without homes, family, food, or any means of support. Their neighbors on the hillsides were the first responders, but major help would soon arrive thanks to rapid repair of the railways as directed by Robert Pitcairn of the PRR, which brought Clara Barton and the Red Cross to the valley.

To understand why the dam breached and destroyed Johnstown and its neighboring boroughs, we have to go back to the beginning. What was the history of this dam? When and why was it built and by whom? What happened to it over the years, and how did the South Fork Fishing & Hunting Club acquire this property? And most importantly, how and why did the Club modify the dam when they repaired earlier damage from 1862? Uldis Kaktins, Stephanie Wojno, and I previously analyzed and compared the South Fork Dam *as originally built*, to the altered dam as modified and repaired by the Club. Fatal changes were made by the Club's workers that doomed the dam to failure (Coleman et al. 2016).

References

ArcGIS (2011) How kriging works. Accessed 11 Apr 2015. Online at: http://help.arcgis.com/en/arcgisdesktop/10.0/help/index.html#//009z00000076000000.htm. Accessed 13 Feb 2016

Blodget L (1890) The floods of Pennsylvania, May 31 and June 1, 1889, pp A143–A149. In Annual report of the Sec. of Internal Affairs of the Commonwealth of Pennsylvania for the year ending November 30, 1889, Part I, 223 p

Coleman NM, Kaktins U, Wojno S (2016) Dam-breach hydrology of the Johnstown flood of 1889 – challenging the findings of the 1891 investigation report. Heliyon 2:54. https://doi.org/10.1016/j.heliyon.2016.e00120

Corning Journal (1889) The Deluge! Thursday Jun 6. Corning J 43(23):3

Dunwoody (1889a) Monthly weather review, vol XVII, No 4, Washington City, May 1889

Dunwoody (1889b) Monthly weather review, vol XVII, No 6, Washington City, Jun 1889

Eng. News (1889) Engineering News. Jun 8 1889

Francis JB, Worthen WE, Becker MJ, Fteley A (1891) Report of the Committee on the cause of the failure of the South Fork dam. ASCE Trans XXIV:431–469

PRR (1889) Statements of employees of the Pennsylvania Railroad Company, and others in reference to the disaster to the passenger trains at Johnstown, taken by John H Hampton, at his office in Pittsburgh, by request of Superintendent Robert Pitcairn; beginning Jul 15 1889

Russell T (1889) The Johnstown flood, *Monthly Weather Review*, May 1889. In US Signal Service Monthly Review, pp 117–119

Shappee ND (1940) A history of Johnstown and the Great Flood of 1889: a study of disaster and rehabilitation [unpublished thesis]. University of Pittsburgh, Pittsburgh

Townsend TF (1890) Pennsylvania State Weather Service Bulletins, pp A34–A142. In: Annual Report of the Sec. of Internal Affairs of the Commonwealth of Pennsylvania for the year ending Nov 30, 1889, Part I, 223 p

USDA (US Department of Agriculture) (1986) Urban hydrology for small watersheds. Tech. Release 55, NRCS, 164 p

Chapter 3
Early History of the South Fork Dam

3.1 The "Age of Canals"

Let us now go back to earlier time in Pennsylvania's history when the South Fork Dam was being planned and built. It was called "The Age of Canals." The Pennsylvania canal system had its origin in the Erie Canal. Construction of the canal in New York State gave a tremendous boost to commerce and development by connecting a water route from New York City and the eastern seaboard to the rich resources of the interior and the lands and lakes farther west. The canal was built during the governorship of DeWitt Clinton, a far-seeing politician and naturalist who had also been a state legislator, U.S. Senator, and three-time Mayor of New York City. In 1812 he was narrowly defeated for the office of President by James Madison. Opened in October of 1825, the Erie Canal was more than 360 miles long, stretching from the upper Hudson in Albany to Lake Erie. Opponents and many newspapers derisively called it Clinton's or DeWitt's "Ditch," but they soon saw the powerful effects of this economic artery on the state's prosperity, including rapidly rising population growth and land values. Clinton lived to see acclaim for his efforts but died suddenly in February, 1828, while still governor. After all he had done for his State and its people, and despite an eloquent state-sponsored funeral, Clinton's family was so bereft of funds they could not obtain his burial plot for years.

The city fathers of Philadelphia did not want to be left behind in this new "Age of Canals." Among these were William Tilghman, Nicholas Biddle, John Sergeant, Mathew Carey, William Lehman, and Benjamin Chew, Jr. Philadelphia had prospered as the leading American city in imports and exports, but by 1825 it was overtaken by New York. Wishing to directly compete with the success of the Erie Canal, political maneuvering began for a trans-state transportation system. In 1824 the Pennsylvania legislature authorized Governor John Shulze to appoint three commissioners to do a feasibility study. Later that year, impatient for faster progress, a group of influential citizens, bankers, and merchants founded "The Pennsylvania Society for the Promotion of Internal Improvements in the Commonwealth." They

© Springer International Publishing AG, part of Springer Nature 2019
N. M. Coleman, *Johnstown's Flood of 1889*,
https://doi.org/10.1007/978-3-319-95216-1_3

published articles supporting a transportation system and encouraged petitions. After receipt of many petitions the legislature crafted a bill that was signed by Governor Shulze.[1] On April 21, 1825 he appointed five members to a Board of Canal Commissioners (Shelling 1938).

That was the beginning, which led to a canal convention in Harrisburg and extensive discussions in the legislature, within the "Improvement Society," in the press, and among the public. It was realized that the project was much too ambitious for private industry or a few wealthy people to sponsor. The Commonwealth would have to authorize and fund the work, a tough task given the precarious state of its treasury. Various issues were raised in opposition. For example, the lucrative state income from turnpike tolls[2] between Philadelphia and Pittsburgh would be lost to a new canal and rail service. How would the work be financed given that the U.S. and the state had lately recovered from heavy expenses and taxation incurred by the War of 1812? And was it even possible to cross the western mountains? In the end, proponents of a Pennsylvania canal led by Philadelphians won the day (Shelling 1938).

During the 1825–1826 session of the legislature, the first report of the Canal Commissioners was received. An act to commence the canal work was introduced and debated. It passed the legislature and was signed by the governor on February 25, 1826. The Main Line of Public Works of canals and associated railways eventually connected the east coast with Hollidaysburg in central Pennsylvania by 1832. West of that place loomed an arresting topographic barrier, the Allegheny Mountains. A railway was needed to ascend and traverse the plateau and ridges, and in 1834 the Allegheny Portage Railroad was completed. To surpass the various crests, the engineers built ten inclined planes, or funiculars (five to the east and five to the west of the mountains), and a mile-long tunnel (*Eng. News*, Apr 20 1889, p 356). This portage railway linked Hollidaysburg with the canal basin and the Conemaugh River in Johnstown, which led to the Ohio River and beyond. The entire region prospered by the commerce between Philadelphia and Pittsburgh and experienced rapid population growth and development along the Main Line.

Place names in Johnstown today reflect the city's canal history. The canal basin with boat docks was north of today's curving Railroad Street, which was formerly a railway, the terminus of the Portage Railroad at the canal basin. Short St. marks the old location of that canal basin. Lock #31 provided boat entry and exit on the west end of the basin. Just west of lock #31, a bridge crossed over the canal at the northern end of Franklin St. The canal itself led northwest from the basin and ran along the north side of today's Washington Street. The canal crossed the Little Conemaugh River on an aqueduct, at or near the site of a present-day steel girder bridge with GPS coordinates 40° 19′ 42.6″, 78° 55′ 09.6″. From that aqueduct the canal led north to lock #30 and then continued northward, east of the old Cambria Iron Works and along the base of Prospect Hill. Present-day Feeder St. was just east of and parallel to an old feeder canal that fed water north from the Stonycreek River into the canal basin.

---

[1] Governor Shulze should also long be remembered for laying the groundwork for free, mandatory education in the state; a feat achieved by his successor.

[2] Those tolls on the modern turnpike are still lucrative today!

3.2   Not Enough Water – A Need for Dams!

The Pennsylvania canals served well enough in seasons with adequate river flows, but by 1834 it was realized that the dry season flows had been underestimated. At such times the rivers could not supply enough water to convey the large number of canal boats. At least 200 and up to 300 boats per day were expected at peak times. If good locations could be found, dams could be built on both sides of the Allegheny Mountains to store water for release during low flows to maintain reliable transport.

In April, 1835 the State Legislature authorized $100,000 for the public works, which included "The commencement of a reservoir near Johnstown." Sylvester Welsh, the principal engineer for the western division of the line, surveyed possible locations for both the Western and Eastern Dams. In 1835 he reported favorable sites for both. For the Western Dam, sites were considered on the Stoney Creek and Little Conemaugh Rivers. An otherwise favorable place on the main stem of the Little Conemaugh was rejected because it would have flooded the village of Jefferson, now known as Wilmore. The watershed of the Stoney Creek was thought to be too large and subject to unmanageable floods.[3] But a favorable site was found on the South Fork of the Little Conemaugh River and was recommended. If a western dam were built at this place it would cover 465 acres and impound 525 million cubic feet of water. A spillway with a bedrock floor would carry water around the end of the dam. No water would be permitted to overtop the embankment. Welsh suggested the dam consist of an embankment with a wall of masonry. The Canal Commissioners were authorized to start building the dams on February 18, 1836. But the work did not begin then.

An 1838 report to the Canal Commissioners asserted that the reservoirs were absolutely necessary and should be built soon. In July the next year the Legislature approved $70,000 to start work on the dams. By July 15, 1839 the sites had been approved by the Canal Board and the work placed in the charge of Principal Engineer William E. Morris. The initial work was under contract by October.

In his November 1839 report to the Commissioners, Morris approved the site for the Western Reservoir recommended by Welsh and discussed the design of the dam to be built. Morris calculated that a dam 62 feet high at this site would be 850 ft. long on its crest and would impound 480 million cubic feet with a lake area of 400 acres. He further reckoned that from storage alone, excluding stream inflows, the reservoir could release 3500 cubic feet per minute for three months to Johnstown via the South Fork and the Little Conemaugh River (Morris 1839).

Prophetically, Morris (1839 p 127) wrote that the site for the western dam "… is situated upon a stream that will furnish an abundance of water to fill the reservoir, and from the floods of which but little danger is to be apprehended, if proper channels [spillways] are constructed for their discharge."

---

[3] The present large dam on Quemahoning Creek was built in the early 1900's.

Figures 3.1 shows the location and extent of the Western Reservoir and the South Fork Dam. The dam was never rebuilt after the 1889 flood. Its remains and the surrounding lands are now part of the Johnstown Flood National Memorial, which was authorized on August 31, 1964. The National Park Service (2018) maintains and protects the property, its roads and trails, a museum, and various historic structures such as the Unger House and the South Fork Fishing & Hunting Club House. An exit from Interstate 219 provides access to the federal park and the villages of St. Michael and Sidman. Many of the houses today in St. Michael were built on the former bed of Lake Conemaugh.

Morris (1839 p 127) further described the site for the South Fork dam:

The valley is narrow at the dam, and widens immediately above, into an extensive basin. The land intended to be flooded, except a few acres, is covered with timber, and consequently but small injury will result to private property. There is solid rock at both ends of the dam, in which channels may be cut for the discharge of waste water, in time of floods. This fact has been satisfactorily ascertained by a full examination, by means of drifts and shafts sunk for the purpose. Abundance of the best material for the formation of the dam, is found convenient. The "feed water" with but little loss from evaporation will pass to the canal at Johnstown, down the natural channel of the stream, which is narrow and protected from the sun by woods and mountains…

…[the dam] to be formed, by a mound or embankment of stone and earth, made perfectly water tight, and raised 10 feet above the surface of the pool – having a waste or channel [spillway] cut in solid rock at one or both ends of the dam, for the passage of the flood water.

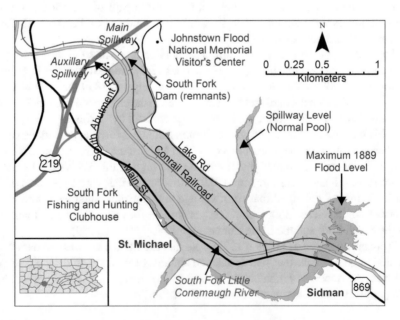

Fig. 3.1 Location map of the South Fork dam and Lake Conemaugh. The towns of St. Michael and Sidman and the Conrail branch line in the old lakebed did not exist at the time of the flood. (Credit: S. Wojno)

Morris estimated a cost for the construction at "...$188,000 00, exclusive of damages, engineering, and office expenses." Morris also projected an excavation volume for the spillways of 40,000 cubic yards.

3.3 Design of the South Fork Dam

William Morris (1841) prepared the initial plans for the Western Dam, its drainage pipes and culvert, and control tower. Those plans were revised and finalized and the contracts awarded on March 10, 1846, for both that dam and the Eastern Dam near Hollidaysburg.

Figure 3.2 shows Morris' final design for the South Fork dam, indicating the slopes and structure of the embankment, the location and form of the control tower, and the position of the drainage (sluice) pipes and the masonry culvert beneath the dam into which the pipes emptied. The maximum height of the dam was 72 ft. (22 m) from its base to the crest, which was 10 ft. wide on top. There were five horizontal drainage pipes made of cast iron, with inside diameters of 24 inches. The contract clearly described the spillway characteristics. A spillway was to be excavated in solid rock in the hill at one or both ends of the dam to discharge flood waters. The combined width of the spillways "...will not be less than 150 feet." While excavating the spillways, a "guard bank" of solid rock was to be left between

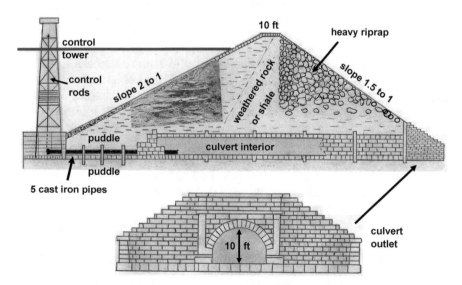

Fig. 3.2 Final design of the South Fork dam, dated March 10, 1846 (after Morris 1846), showing heavy riprap on the downstream face, a broad "puddled" clay section on the lake side, slope wall of dressed stone on part of upstream face, and masonry culvert beneath dam, fed by cast-iron pipes to discharge water and control lake level. The pipes led from control tower to upstream end of culvert. Embankment was 72 ft. high

the ends of the dam and the inner slope of the spillways. The ends of the spillways had to be at least 50 feet beyond the outer slope of the dam. This last specification was to ensure that water flowing through the spillways would never erode and destabilize the downstream base of the dam.

Flow to the discharge pipes would be controlled using stopcocks operated by long rods from the top of the wooden tower in the lake near the dam center. By design the "sluices" were actually trains of cast-iron pipe sections joined together at bell sockets sealed by pouring molten lead into the joints. There were five "trains" of pipe, each consisting of 12 individual 7-ft (2-m) segments. Due to overlap in the socket the effective length of each pipe segment was 6 ft. 5.5in (Unrau 1980 Appx. K), and the total length was about ~23.6 m (77.4 ft). By comparison, the pipes for the Eastern Dam at Hollidaysburg, which had a lower embankment, were shorter with a total length of 70.4 ft. The pipes were specified to have an "in the bore" [inside] diameter of 24 in (0.61 m) and had to undergo integrity tests "joint by joint" using a force pump up to a pressure of 300 ft. of water (900 kilopascal).

3.4 Building the Dam

Four rich sources of information about the South Fork dam include the annual reports of the Canal Commissioners, the investigation report by Francis et al. (1891), the 1940 doctoral dissertation by Nathan Shappee, and field books recorded by the engineers in charge of construction, now kept at the Pennsylvania Archive in Harrisburg.

The property for the Western Dam was eventually acquired. The construction contract was placed for public bid and awarded to James Morehead & Co. on November 6, 1839. The Canal Commissioners gave their approval in December and David Watson was chosen superintendent of the reservoir. The contract was dated January 31, 1840 with contractors James K. Morehead and H. B. Packer (Francis et al. 1891). The contract specified prices to be paid for various works. For example, digging of "slate" in the spillways would be reimbursed at 32 cents per cubic yard. "Slate" was a term then used for degraded rock that could be removed with a pick and shovel. The contractors would be paid 34.5 cents per cubic yard to excavate "solid rock" from the spillways. The estimated volume to be excavated from the spillways had now increased to 50,000 cubic yards.

When eventually filled the impoundment would inundate lands owned by George Murray, Joseph Leckey, Philip Myers, Daniel Bumgardner, and Est Conrod Fye. More than 300 acres of Murray's property would be covered by the lake. The work began with grubbing and clearing in the spring of 1840. This was brutal work, removing all vegetation from the 400 acres to be flooded, nearly all of which was wooded. If not removed the vegetation would rot in place beneath the lake and some could eventually detach and drift along the lake bottom down to the dam, clogging the screen that covered the intakes of the discharge pipes. That screen would be impractical or impossible to unclog once the lake had formed. Useful timber was cut down and sold to local buyers, or taken by wagon to load on railroad flatcars. Tree tops were pulled out of the basin by horse teams. Large shrubs were yanked out by their roots using ropes and chains hitched to horses and mules. Material not useful for timber was burned. No degradable organic matter was to be placed in the embankment.

As the future lake bed was being cleared of organic debris the workers also turned their attention to the embankment foundation. Here too, all vegetation and sod had to be completely removed. "Puddle" ditches were then dug to a depth specified by the engineer in charge, which might include some depth into the bedrock. The embankment would then to be raised according to Morris' plans such that the dam crest would be 10 ft. wide on top and reach a height 10 ft. above the water line, which represented the spillway floors or "normal pool" for the lake surface.

By late 1840 most of the clearing and grubbing had been done and the cast iron drainage pipes fabricated and ready for transport to the site. The foundation for the masonry culvert had been laid, but the arched culvert walls were not yet built. "Puddle" ditches were cut ~5 feet below the base of the pipes. These ditches made it possible to surround the pipes with clay layers of low permeability, thus minimizing water leakage from the reservoir through the lower part of the dam. This approach stabilized the dam by preventing fine materials from washing out through the embankment. By November 30, 1841 Morris estimated that the cost of the work done so far under the contract was $80,000. He gave the following report:

> Since last fall the contractors have steadily pushed on the work at the dam, though, from the smallness of the appropriation, with a moderate force. The sluice walls are raised sufficiently high to receive the pipes, each range of pipe about 80 feet long, has been laid and tested by a head of 300 feet.

Unfortunately Pennsylvania's finances were in such bad shape that money for public works had to be cut to bare necessities. An act was passed on April 3, 1841 authorizing $50,000 for all the work done through May 1st of that year at the Western and Eastern Dams. Then work on both projects stopped (Francis et al. 1891).

A long hiatus ensued, during which parts of the embankment washed away. Five years later work resumed at the South Fork dam after the legislature approved $20,000 for that purpose on January 3rd, 1846. The Canal Commissioners were tasked to complete the dam quickly. William Morris provided new plans for the dam, and these were exhibited on March 3rd of that year. These are the plans shown in Fig. 3.2 above, which bear the date March 10th, 1846. The only change from his 1841 design was to save costs by building the control tower of wood rather than masonry. However, the dam work was postponed yet again because a spring flood caused damages to the canal system elsewhere that had to be repaired immediately (Francis et al. 1891).

Another four years went by before a new appropriation was made, but further delay caused those funds to return to the general State fund. Finally, on April 15, 1851 the legislature authorized $45,000, and work resumed on the South Fork dam two weeks later. The embankment grew rapidly that year. An additional $55,000 was provided on May 4th, 1852 to complete the dam. Six weeks later, on June 10th, the discharge pipes were closed and the basin began to fill. The water slowly rose so that by early September the lake was 40 feet deep near the dam. The season was quite dry that year, and water had to be released to support the canal transport. All work was finished by 1853 (Francis et al. 1891).

Because knowledgeable engineers designed and oversaw construction of the South Fork Dam, it was decided not to allow the lake depth to exceed 50 feet during 1853. That decision was prudent then and likewise a sage one for modern dams with new embankments. Any problems found in a new embankment are easier to correct if the lake behind it is not full. This lesson has been learned the hard way even in modern times. The Teton Dam in Idaho failed in 1976 while being filled, although the root causes of its piping failure related more to a design that did not fully consider geological factors related to fill materials and the foundation of the dam (Chadwick et al. 1976). At least 11 people perished along with more than 10,000 cattle. If the dam had been filled more slowly it might have been possible to identify hazardous issues earlier, with less water in the reservoir.

It is not well known that the reliability of the South Fork dam was tested just a few years after it was completed, when in the spring of 1856 the reservoir overflowed at an unspecified place after a rapid snowmelt in March (Unrau 1980 p 51). It was not reported where the overflow occurred, but given the higher dam crest at that time it could only have flowed over the western abutment, which was lower than the embankment and served as an auxiliary spillway. An overflow about one meter deep could have surged over the western abutment without overtopping the embankment and without damaging the toe of the dam. The reservoir clearly performed as intended with two functioning spillways and the discharge pipes also probably opened. A leak in the dam was reported at that time but was soon repaired. In fact, the Pennsylvania State Engineer inspected and reported about the South Fork dam later that year. "The Western Reservoir [Lake Conemaugh] was examined and found to be in excellent condition. It furnished a sufficient supply of water to keep up the [canal] navigation when other sources had entirely failed" (Gay 1856 p16). He also reported that the spring floods severely damaged other impoundments in the region, including Piper's and Raystown dams on the Juniata River. Eighty-foot breaches occurred in both dams.

3.4.1   Excavations for Dam Material

The builders of the original South Fork dam got their material from nearby sources. In fact specifications attached to the State contract required that material be obtained within ¼ mile of the site (Francis et al. 1891). There were three main sources: earth and rock taken from the spillways; quarries on the hillside above the dam; and earth and rock removed from the valley floor upstream from the dam, producing terraces that are visible today. All organic debris was to be excluded.

New and detailed information is now available about how the South Fork Dam was built. Thanks to LiDAR mapping we can identify where the material was quarried to build the dam and cover it with riprap. "LiDAR" (a portmanteau word combining "light" and "radar") is a remote sensing technique that measures distances very accurately using lasers and sensitive optics. Point data are obtained with x, y, and z values that specify each point's lateral position and elevation. The data are processed to remove buildings, canopies of trees, and other features to generate a "bare Earth" relief map that can be converted to a digital elevation model (DEM). LiDAR data have been available for all of Pennsylvania since 2008. The data allow land surface mapping in much finer detail than was possible with previously available topographic maps and DEMs of lower resolution. Reports of vertical accuracy testing for Pennsylvania's LiDAR datasets are provided by DCNR (2018).

My colleague, Stephanie Wojno (and co-author of Chap. 13), provided LiDAR maps of the main spillway (see Appendix 2) and the western abutment at the South Fork Dam. The dam builders excavated an auxiliary spillway by terracing the western abutment of the dam where a parking lot now exists. They also dug quarries on the hillside and on the floor of the former lake bed (see arrows, top frame, Fig. 3.3). Excavations in the valley south of the dam, below the expected lake level, had the great advantage of increasing the storage capacity of the lake. Long terraces were created by this work in the future lake basin (Fig. 3.3).

Figure 3.4 is a post-flood image from 1889 showing laminations from "puddle ditches" in the exposed side of the breach. The puddle technique was used to reduce infiltration of lake water into the embankment. Note the utility building at the upper right corner of Fig. 3.4. This structure was built near a level, excavated terrace that was created to provide an auxiliary spillway to protect the dam at times of high lake stands.

The contractors who built the original dam were required to obtain their materials from places near the site. No such restriction existed for the South Fork Fishing & Hunting Club, but obviously the closer the quarries were to the dam the quicker the mined material could be brought and placed in the embankment. They did incorporate organic material, which an experienced hydraulic engineer would not have done.

Because coal mines were operating in South Fork at the time the Club repaired the dam, Kaktins et al. (2013) suggested that some mining waste may have been placed in the embankment. Also, "sulfur water" had been reported seeping from the embankment.

Accounts stated that the fill included dirt, clay, shale, old bricks, and, once again, brush, hay, and straw. It is probable that at least part of the fill came from cheap and easily attainable coal- and clay-mining wastes. The Lower Kittanning coal crops out in the South Fork area and was extensively mined at this time. A plastic clay is associated with the Lower Kittanning seam and this clay would have been in the waste rock from mining operations. One of the club members, Pitcairn, had a friendship with the superintendent of the Argyle coal company… [Kaktins et al. 2013 p 342–343]

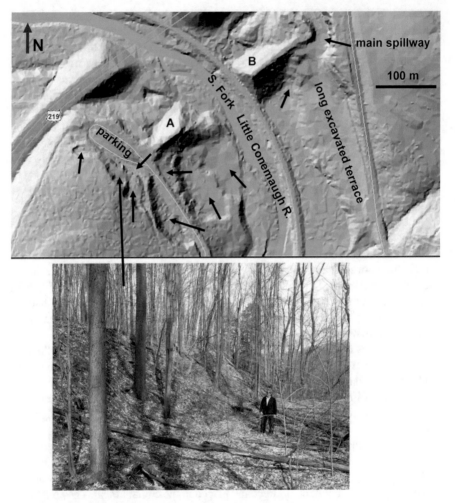

Fig. 3.3 Top: LiDAR image of remnants (A and B) of South Fork Dam and part of the old lake basin. Black arrows point to quarries where soil, regolith, and bedrock were dug and quarried to obtain embankment material and create both spillways. The excavations produced terraces in the present landscape. Modern roads are shown. Loop at left center is western parking lot of Johnstown Flood Memorial. Auxiliary spillway was at southeast end of parking lot (base image credit: S. Wojno). Bottom: Quarry on hillside south of parking lot (location shown by long arrow). Photo 2017 by the author

No coal mines existed near the dam when it was first built, so mine refuse would not have been available for the original embankment.[4] I found no evidence to differentiate where the material was obtained for the original dam vs. the repair

---

[4]The Maryland Shaft #1, a deep mine at St. Michael, did not start operations until circa 1908.

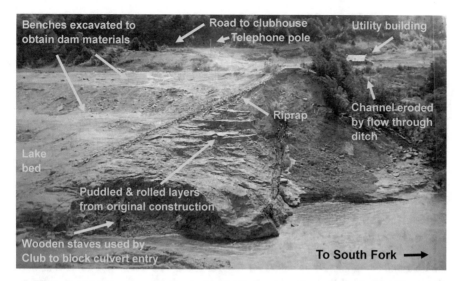

Fig. 3.4 Post-flood features of western dam remnant. Note excavated benches on the upstream side of the dam where material was obtained for the embankment. These deepened the lake bed and increased the lake volume. This view of the western abutment (*upper right corner*) shows that it was clearly excavated to a terrace, forming the second spillway, but apparently due to cost cutting measures it was only dug to a depth of 3 to 4 ft. Base image courtesy of the Johnstown Flood Museum Archives, Johnstown Area Heritage Association

work in 1880–81 by the South Fork Fishing & Hunting Club. The Argyle Mine[5] existed in 1881, but it seems unlikely the Club workers would have transported thousands of cubic yards of mine waste by horse-drawn wagon, uphill a distance of two miles from the mine area to the dam. Common sense and time efficiency dictated they seek fill near the dam. But given free access to easily loaded waste, it is possible some mine material was sometimes brought from South Fork rather than drive empty wagons up to the dam. Along with the clay fraction naturally present in soils near the dam, the placement of some wagon loads of mine waste might have added enough plastic clay to affect the stability of the dam on the day of the flood.

---

[5] Inspector Roger Hampson reported on the Argyle Mine (probably #1) in 1882 and found it to be "in very poor condition …" Ventilation was the concern. J. P. Wilson was the mine superintendent. Fifty men worked there as of 1882. On Feb. 8th that year a 20 year old named D. O. Kuhns was instantly killed at the mine in the presence of Wilson. He had fallen and a loaded rail car rolled over his neck. Two weeks later Simon Richardson, age 21, died in a fall of coal. At that time there were only four bituminous mine inspectors in the state; James Louttit, John Davis, Thomas Adams, and Roger Hampson, whose territory included Cambria County. Hampson (1883) reported that 1463 persons worked in Cambria County to produce coal and coke, as of the end of 1882.

3.5 Decommissioning the South Fork Dam

Two minor leaks in the South Fork Dam occurred in 1854 but these were quickly repaired. After its long history of construction delays and considerable cost, the dam served the canal system for only three more years. Along with the Portage Railroad it was already obsolete! New rail lines had been surveyed across Pennsylvania, and advances in locomotive engines allowed direct rail transport across the State without the need for canals or the awkward system of inclined planes on the Portage Railroad. One of the new routes included the famous "Horseshoe Curve" near Altoona. In 1857 the State of Pennsylvania sold the entire system of the Main Line Canal to the Pennsylvania Railroad Company. The sale included the South Fork Dam and its "sister," the Eastern Dam near Hollidaysburg. After that time the canal system was still used for local service but was no longer a primary mode of transporting goods across the State. The dams were not needed by the railroad and were largely neglected with minimal oversight. Mr. Joseph Leckey, who lived near the South Fork Dam, served as caretaker for the PRR. The lakes still existed at both dams and were used by locals for fishing and recreation, to the extent permitted by the railroad's watchmen.

3.6 Partial Breach of the Dam in 1862

It is not widely known that the South Fork dam had partly breached many years before the 1889 flood. Sometime between 1857 and 1862, leaks had been reported beneath the dam where the five drainage pipes entered the stone arch culvert at the base of the dam (Francis et al. 1891 p 445). In mid-July of 1862 the culvert partially collapsed. "A portion of the arch in the breast wall has fallen ... Should the dam give way suddenly... unless the fallen wall is speedily repaired, the consequences would be serious." The matter was brought to the attention of Borough Council (*Cambria Tribune* 1862a Jul 18). Less than two weeks later, on July 26, a breach formed that drained the lake (*Cambria Tribune* 1862b Aug 1):

> The Reservoir dam, the precarious condition of which we noticed two weeks ago, gave way on Saturday morning last, and emptied its waters into the Conemaugh [River]. The announcement of the breaking of the dam caused considerable alarm in town, but owing to the low stage of water in the creek the flow from the Reservoir produced but an inconsiderable rise, and the excitement and the flood both soon subsided. No loss or damage was sustained by anybody so far as we can learn, except the carrying away of about $200 worth of bridge timber belonging to Wood, Morrell & Co., which was being floated down the creek, and the overflowing and washing away of a few rods of the railroad track at South Fork, which detained the morning train from the East until late in the afternoon. Many people were badly scared about the breaking of the Reservoir, but nobody was hurt by it.

As it came about, caretaker Leckey saw muddy water flowing from the culvert and assumed the upstream end was further collapsing, which it had been threatening to do. He climbed into his boat and rowed to the wooden control tower, then fully opened the stopcocks for the discharge pipes. Leckey then quickly saddled a horse and rode to Wilmore, 6 km from South Fork (the South Fork station was not built until 1883). George Kerbey was ticket agent and telegraph operator for the Pennsylvania Railroad at Wilmore; he sent a warning to Johnstown about the dam (*Johnstown Daily Tribune* 1900).

The breach took half a day to fully form and this slowed the outflow considerably. The manner of construction of the dam, with many layers of "puddled clay" in the embankment, slowed the rate of down-cutting. Leckey's action to open the pipes may have lowered the lake level more quickly, but might also have sped up the culvert collapse. Fortunately the lake was low at the time, only about 48 feet (15 m) deep at the dam. At that depth the impoundment had only half the storage contained when the lake was full. The embankment was nonetheless seriously damaged by the partial breach. The final collapse and flood erosion washed out a triangular section of embankment about 60 m wide. A sawmill and small house below the dam were washed away, in addition to the rail washout farther downstream.

Shappee (1940 p 211) gives one idea about how the dam might have been damaged, leading to the 1862 breach. The railroad's watchman sometimes had to drive away local neighbors who coveted the lead used in the joints and fittings of the sluice pipes. Some of those pipe fittings were exposed at the upstream end of the culvert. Theft of lead would have been possible because people could enter the culvert through its lower end, and the inside space from the floor to the top of the arch was 10 ft. high (3 m). The plans for the dam had specified that "A suspension way or walk will be constructed from the lower end of the culvert to the stop-cocks at the end of the pipes; it will consist of a plankway 2½ feet wide, and 2 inches thick, suspended from the arch of the culvert by iron rods and terminating upon the culvert wing." (Francis et al. 1891, p. 443). Although there is reference here to stopcocks at the end of the pipes, the revised plans in 1846 seem to show stopcocks only at the tower base, the upstream entry of the pipes. The suspended walkway was intended to allow access for direct inspection of the culvert interior and pipe ends even when some flow was occurring through the pipes.

The five large drainage pipes protruded into the upstream end of the culvert and would likely have been laid on lead strips or fittings used to seal the annular spaces between the pipes and the masonry supports. It would not have been difficult for thieves to evade the watchman. Simple tools would have sufficed to pry out chunks of lead which as scrap could be used in various ways.[6] Removal of the lead fittings and seals would have allowed water to leak into the culvert through the "puddled" backfill that surrounded the iron pipes. As this flow increased over time it would have gradually piped material out of the embankment and into the drainage culvert.

---

[6] Lead would have been more costly in 1861 and 1862, the initial years of the Civil War. The average price of pig lead in New York during 1853–1862 was 6.13 cents/pound (Congress, 1913), but retail prices would have been higher. Average prices diminished after 1872. But "free" lead would always have been desired by those with little cash, such as young lads wanting to melt it down to make fishing weights or bullets for hunting.

Eventually enough material could wash out to destabilize the upper end of the culvert. And that is where the collapse occurred, leading to the 1862 breach.

It is ironic to contemplate how a seemingly minor pilfering of lead, if true, could have begun a fatal chain of events, starting with early damage to the culvert beneath the dam, leading to the 1862 partial break, inadequate repair and other changes by the Club, and eventual destruction by dam breach of the downstream towns.

3.6.1 Possible Role of Hydrodynamic Damage or Other Processes

Some damage to the culvert and cast iron pipes could also have been caused by hydrodynamic effects such as cavitation, during the years the dam supplied water. I analyzed flows in the discharge pipes to see if they were energetic enough to start and sustain cavitation damage, given the presence of heterogeneous surface roughness (Coleman et al. 2016). The inception of cavitation and damage from bubble-collapse shock waves involve complex flow dynamics, which are not easy to quantify. Cavitation can begin when the local fluid pressure falls sufficiently below the equilibrium (or saturated) vapor pressure. Falvey (1990) gives a detailed discussion of the conditions and mechanisms, including the effects of different kinds of surface roughness (i.e., grooves and voids, protruding joints, and various offsets). Based on the construction methods used, cavitation inception would have been enhanced in two zones beneath the South Fork dam: (1) inside pipes near the boundary zones between pipe sections where small offsets would have existed; and (2) at the upstream end of the masonry culvert, where the sluice pipes discharged into the culvert. Damage from vapor bubble implosion always occurs downstream from the source of the cavitation. Cavitation forms by turbulence in a shear zone, resulting from sudden changes in flow direction at irregularities. Any space or irregularity in the seam between two pipe sections could have initiated cavitation, given sufficient flow speeds.

The construction method used for the sluice pipes in the South Fork dam was well documented. Each range of pipe consisted of 12 sections of cast iron pipe, each ~2.1 m long that were joined together by lead fittings at bell-shaped socket joints (Unrau 1980 p 22, 27) and laid horizontally, emptying into a chamber beneath the dam at the upstream end of the culvert. Thus, each 23.5 m length of pipe would have had 11 "rings" of potential cavitation zones, where shock wave damage would be focused on the pipe walls near sealed joints. Cast iron is hard but very brittle, and could readily have been eroded by vapor bubble collapse downstream from these roughness elements. During construction the pipe joints had been tested to withstand a pressure equal to 90 m of water pressure. An analytical equation for the inflow capacity of a pipe spillway at full flow (Chow 1964) yields estimated average pipe flow speeds of >13 m s^{-1}, which under a lake pressure head of 21 m would indicate an incipient cavitation index of >2.2 (Falvey 1990 Fig. 2-3a). This result suggests cavitation damage was unlikely in the South Fork drainage pipes. The same conclusion is found with the methodology of Hancock et al. (1998 Eq. 12). However, in subsequent research, the same authors (Whipple et al. 2000 p 498) conclude that for flows with high Reynolds numbers[7] (10^5–10^6) cavitation can commonly occur at cavitation indices of 2–3. Cavitation damage would thus have been possible, but only for limited times at the highest lake levels when the pipe flows could achieve Reynolds numbers reaching 6×10^6.

I also examined the conditions that could have caused hydraulic damage in the upstream culvert chamber where the iron pipes emptied. This was the part of the culvert that collapsed in 1862. Abrasion from suspended load would not have been significant because the upstream ends of the pipes were open to the water column of a lake. Calculations of open-channel flow based on a Q of 20 m^3 s^{-1}, flow width of 3.4 m, flow depth of 1.5 m, culvert floor slope of 0.0055, and Manning $\eta = 0.016$ indicate a mean flow speed <4 m s^{-1} (for critical flow; Froude # = 1). Cavitation was unlikely at this flow depth and speed, and greater η values would further reduce the speed. Deeper flow was needed, and mean flow speeds would have had to exceed ~9 m s^{-1}, to support cavitation inception. However, the presence of five adjacent pipes discharging jets of water into the masonry-lined chamber would have generated complex turbulent flow with strong shear forces, spawning vortices that would have imposed lifting forces on inner surfaces of the culvert. Vortices may also have raised local flow speeds near asperities enough to approach cavitation conditions. The energetic flows would have weakened some of the mortar joints over time, allowing material to be piped from behind the masonry work and destabilizing the supporting walls of the arched culvert. If these processes were responsible, then most of the hydraulic damage likely occurred at times when the lake was mostly full and the pipes were discharging to augment flow in the canal system. High lake levels would have maximized flow speeds in the pipes and culvert.

It was also considered whether winter cold could have penetrated beneath the dam from the open outlet to the upstream end of the culvert where the pipes protruded. In such a case frost shattering might have contributed to damage of the culvert lining prior to 1862. However, the distance from the outlet to the upper end of the culvert was about 60 m. This distance, combined with conductive ground warmth beneath the dam and additional warmth from any water seeping through the pipes, would have protected against freezing at the upstream end. By analogy, one does not have to go far into a cave or mine in winter to find above-freezing conditions, especially in a dead-end drift with no through-flow of air.

And finally, there is a possibility that the pipe seals and culvert joints could have been damaged by slight differential settling of the dam, given that the pipes lay beneath the deepest fill at the embankment center.

The culvert and discharge pipes are gone, so it is impossible to further study the cause of the 1862 partial breach. However, the analysis presented here suggests that cavitation and other forms of hydrodynamic damage may have contributed to the

---

[7] Reynolds number is a dimensionless ratio of inertial forces to viscous forces; at sufficiently high values, turbulent flow dominates over smooth, laminar flow. Large Reynolds numbers are needed to induce cavitation.

partial dam failure. The details of cavitation were unknown to science and engineering when the dam was built, but its design did not make the dam highly susceptible to cavitation damage. Damage was possible during pipe discharges at high lake levels. Theft of lead seals and fittings at the upper end of the culvert would have been tempting and easy to do and would have allowed seepage into the culvert from the embankment. Over time, gradual piping of overlying material would have occurred, destabilizing the culvert and eventually causing the muddy water issuing from the dam, as seen by the caretaker just before the collapse and partial breach.

In any event, the large Western Reservoir was now gone. A small lake remained near the dam to be visited by local fishermen; a watchman was no longer needed. The Pennsylvania Railroad kept the property for another 13 years, when in 1875 they sold it to one of their own employees, John Reilly of Altoona, who had resigned to serve his term as a representative in the U.S. Congress. And there begins the story of the South Fork Fishing & Hunting Club.

References

Cambria Tribune (1862a) [No title – item under local news]. Bowman J (Ed), Vol 5, no. XVI, July 18, Johnstown, PA

Cambria Tribune (1862b). [No title – 1st item under local news]. Bowman J (Ed), Vol 5, Aug 1, Johnstown, PA

Chow VT (ed) (1964) Handbook of applied hydrology. McGraw-Hill Book Co., NY, p 1418

Coleman NM, Kaktins U, Wojno S (2016). Dam-breach hydrology of the Johnstown flood of 1889 – challenging the findings of the 1891 investigation report. Heliyon 2 (2016), p 54. https://doi.org/10.1016/j.heliyon.2016.e00120 [cavitation discussion in supplementary online material]

Chadwick et al. (1976). Report to US Department of Interior and State of Idaho on Failure of Teton Dam, by Independent Panel to Review Cause of Teton Dam Failure Dec 1976, p 590

DCNR (2018) PA Department of Conservation and Natural Resources, link to vertical accuracy reports for LiDAR data in Pennsylvania. Available at: http://www.dcnr.state.pa.us/topogeo/pamap/documents/index.htm. Accessed 4 May 2018.

Eng. News (1889) Engineering News, Apr 20 p 356

Falvey HT (1990) Cavitation in chutes and spillways, engineering mono. No. 42, Bureau of Reclam. p 145

Francis JB, Worthen WE, Becker MJ, Fteley A (1891) Report of the committee on the cause of the failure of the South Fork dam. *ASCE Trans* XXIV:431–469

Gay EF (1856) Report of the [Pennsylvania] State Engineer for the fiscal year ending Nov. 30, 1856. Report dated Dec 23, 1856. In: Reports of the heads of departments. Transmitted to Governor in pursuance of law, for financial year ending Nov 30 1856. Harrisburg

Hampson R (1883) Mine and accident inspector report, 4th bituminous district, dated Feb 9, 1883; In: Annual Rept of the Sec. of Internal Affairs of the Commonwealth of Pennsylvania, Part III, Industrial Statistics, Vol X, 1881–82

Hancock GS, Anderson RS, Whipple KX (1998) Beyond power: Bedrock river incision process and form, in rivers over rock: fluvial processes in bedrock channels. Tinkler K J and Wohl E E eds), AGU, Washington, DC, 35–60. doi: https://doi.org/10.1029/GM107p0035

Johnstown Daily Tribune (1900) The water problem. Swank G T (Ed), Aug 10, Johnstown, PA

Kaktins U, Davis Todd C, Wojno S, Coleman NM (2013) Revisiting the timing and events leading to and causing the Johnstown flood of 1889. PA History: A Journal of Mid-Atlantic Studies 80(3):335–363

Morris WE (1839) Report of William E. Morris, engineer [No. 12]. Reservoirs. Engineer's Office, Hollidaysburg, Nov. 1, 1839. To: James Clarke, Esq., President of the Canal Board, pp 125–131

Morris WE (1841) Reservoirs, report of William E. Morris, engineer, Nov 20, 1840. PA House Jour., Appx to Vol 2, p 401–405

Morris WE (1846) PA State Archives, PA Canal Maps, map book #4, page 11, plan dated Mar 10 1846, available at: http://www.phmc.state.pa.us/bah/dam/rg/di/r017_0452_CanalMapBooks/ CanalMapBook4Interface.html. Accessed 27 March 2018

NPS (2018) Johnstown Flood National Memorial, National Park Service, website: https://www. nps.gov/jofl/index.htm. Accessed 9 April 2018]

Shappee ND (1940) A history of Johnstown and the great flood of 1889: a study of disaster and rehabilitation [unpublished thesis]: University of Pittsburgh, Pittsburgh, PA

Shelling RI (1938) Philadelphia and the agitation in 1825 for the Pennsylvania Canal. Pennsylvania Magazine, LXII (Apr 1938) 175–204

Unrau HD (1980) Historic structure report: the South Fork dam historical data, Johnstown flood National Memorial, PA. Package no. 124, US Department of Interior, NPS (1980): 242

Whipple KX, Hancock GS, Anderson RS (2000) River incision into bedrock: mechanics and relative efficacy of plucking, abrasion, and cavitation. GSA Bull 112:490–503

Chapter 4
The South Fork Fishing & Hunting Club

On Wednesday, October 15th, 1879 the *Johnstown Daily Tribune* printed an article titled "A New Summer Resort." The story said that the Western Game and Fish Association of Pittsburgh had bought the property and wanted to *immediately* "… go to work on the dam of the old Reservoir, on the South Fork of the Conemaugh River… It is the intention of the corporation named to commence rebuilding the dam and putting the extensive grounds in proper shape for the erection of a summer resort…" But unknown to the newspaper editor, the article contained a lie – the Association had not yet bought the land and would not own it for another 5 months.

The same edition of the paper carried a small ad asking for 50 men to work on the South Fork Dam (Fig. 4.1). The contractor named was Daniel Kaine. Substantial work was indeed done on dam repairs that late fall. But then in December a week with heavy rains brought snow-melt flooding to the South Fork of the Little Conemaugh. All the material that had been dumped into the dam breach was washed out on Christmas day. This was reported in the *Tribune* (1879b), which also repeated the local belief that the Club had bought the old dam several months earlier. The Club workers decided to delay more work until the next summer. Resuming work in the spring would have been counterproductive with frequent rains keeping the streams high.

After all these years it is still hard to understand the secrecy behind the name and formation of the Club. Even without owning land the Club was incorporated, in Pittsburgh rather than Johnstown, in May 1879. Its charter was approved late that year by a Judge in Allegheny County, and today few people know that the official name was the South Fork Fishing and Hunting Club of Pittsburgh. Pittsburgh was listed as the Club's official place of business. No doubt the charter was never in jeopardy given the 16 names listed, which included Henry Clay Frick and Christopher Hussey[1] and other prominent business people of Pittsburgh (Connelly and Jenks 1889). Magnates like Frick, long associated with Andrew Carnegie, were

---

[1] Hussey was deceased before the 1889 flood.

© Springer International Publishing AG, part of Springer Nature 2019
N. M. Coleman, *Johnstown's Flood of 1889*,
https://doi.org/10.1007/978-3-319-95216-1_4

Fig. 4.1 Advertisement in
the *Tribune* (1879a) for 50
men to work at the old dam

accustomed to keeping their future business plans under wraps, as future profits often depended on early favorable decisions and acquisitions. And of course the elite, then as now, placed great value on the privacy of their personal lives.

4.1   How the Property Was Acquired by the Club

But the question remains – why was extensive work being done at the old South Fork dam by a large crew on land the Club did not yet own?? This early dam work by the Club relates to scrap iron. Between 60 and 70 tons of cast iron existed in the five large drainage pipes beneath the South Fork dam, and these pipes were removed before the Club workers repaired the dam. There are two stories about who pulled the pipes. In one version the Club removed them and in the other it was the land owner, former Congressman John Reilly. Enough information is now known to answer this question, and the answer differs from any previously published. To understand who removed the pipes we have to go back in time and review more history of the dam and the Club, the land ownership by Reilly, and the timing of land transfers and work done at the property.

After the Pennsylvania Railroad acquired the property from the State, the dam breached gradually in 1862 and drained the lake in half a day. The dam was not repaired because the Pennsylvania Railroad no longer needed it having abandoned the Main Line Canal system. The PRR held on to the property for 13 years. In 1875 the Company sold[2] a 500-acre land parcel, including the dam and former lake basin, to John Reilly, who had just been elected to the 44th Congress. Reilly resigned from the PRR to begin his Congressional term, which would start in March, 1875. Before the month was out he was the new owner of the South Fork dam, bought from his former employer, and with his new prominence in politics he no doubt got a "sweetheart" deal from them. The PRR had great influence in politics, both within the State and beyond, and would have curried Reilly's favor.

After serving his term in Congress (March 4, 1875 to March 3, 1877) and losing reelection, John Reilly returned to Altoona in 1877 to work for his old employer as superintendent of transportation. Unrau (1980 p 59) claims that Reilly had failing

---

[2] Pennsylvania Railroad Company to John Reilly, March 29, 1875, Cambria County deed book, vol. 38, 56–58.

health in 1879 and therefore planned to sell the South Fork dam property before leaving Altoona. I found no evidence that he was seriously ill, in fact he lived for another 24 years. Reilly probably just wanted to unload his mountain real estate before leaving for the big city. In 1881 his PRR office moved to Philadelphia and he went with the job, taking his family. Reilly became general superintendent for the PRR and remained in that job until 1885, when his life truly was at risk after a serious accident. He resigned from the railroad but eventually recovered his health. He was involved with various businesses in Philadelphia, where he became president of the United States Metallic Packing Company, vice-president of the American Locomotive Sander Company, and president of the Millwood Coal and Coke Company. These and other details about Reilly[3] and his family are provided by Jordan (1921), who includes a portrait of the former Congressman.

In 1879 Reilly came to an agreement with Benjamin Ruff to transfer the property. Ruff wanted to repair the South Fork dam and establish a resort. As already noted, intensive work began at the old dam 5 months *before* Reilly transferred the real estate, and 1 month before the charter was approved for the South Fork Fishing & Hunting Club. Reilly eventually sold the property, legally transferring it to the Association on March 15, 1880 (LDA Architects et al. 1993 p 437):

> [Deed, John Reilly, et ux, to South Fork Fishing and Hunting Club. 3/15/1880, Cambria County Deed Book 4: 319–322. 500 acres, 54 perches; $2000]

4.2 So Who Pulled the Pipes?

I now return to the two versions of the story about who removed the discharge pipes. The ASCE investigators (Francis et al. 1891) wrote in their report that the Club removed the pipes. They had interviewed local citizens, neighbors of the dam, and likely learned about the fate of the pipes from long-term neighbors who had witnessed happenings at the dam over the years. Certainly the new work of a 50-man crew would have drawn everyone's attention. But the ASCE investigators did not identify who they talked with and what they learned from each person. Something like the Pennsylvania Railroad (PRR) interviews would have been helpful to document the history. Francis et al. (1891 p 445) wrote that the dam repair was contracted out to Colonel B. F. Ruff, who in reality was the promoter and first president of the Club. It is most likely that the Daniel Kaine mentioned in the *Tribune* advertisement worked for him, and would have had to deal with the pipes from the outset.

---

[3] Reilly passed away at his home on April 19, 1904 at age 69. He and at least 4 family members, including wife Anna Elizabeth [Lloyd] and son John Reilly Jr., were interred at West Laurel Hill Cemetery in Philadelphia, in a large plot with a single central monument. There are no individual markers. Some prominent families chose this out of fear that their graves, if well marked, might be targeted for theft or desecration. It was well known that in 1876 two people tried but failed to steal Abraham Lincoln's body and keep it for ransom.

Robert Pitcairn, in his PRR (1889) testimony in July, claimed that John Reilly removed the pipes:

Q.   [Hampton] *Did the Pennsylvania Railroad Company ever rebuild that [South Fork] dam?*
A.   [Pitcairn] No, sir.
Q.   *What became of the property after the dam broke [in 1862]?*
A.   [Pitcairn] It was sold to John Reilly. He took up the pipe, for the purpose of which he bought it, and sold the pipe.
Q.   *What pipe do you allude to?*
A.   [Pitcairn] The pipe that conveyed the water from the dam to Johnstown to the canal.

As a member of the Club and as a long-standing railroad employee and executive, Pitcairn should have known the truth about the iron pipes. But, and this is important, by the time of the PRR testimony it was well known and discussed in the newspapers that the pipes should either have been kept or replaced to allow control of the lake level and thereby permit repairs to the South Fork Dam. Robert Fulton, an engineer and geologist who had inspected the dam in November 1880, had emphasized this. Therefore, Pitcairn would have been motivated to skew history a bit by claiming that Reilly pulled the discharge pipes rather than the Club. I believe Pitcairn, in the flood aftermath, wanted the written record of the PRR testimony, *which he controlled*, to reflect that Reilly was the one who removed the pipes.

In a recent work, Kooser (2013) concludes that John Reilly removed the pipes because Robert Fulton's inspection report noted the lack of a discharge pipe to control the lake level. However, Kooser did not consider that Fulton visited the dam *a full year after* the 50-man crew began its work. At the time of Fulton's visit the repairs to the embankment had far progressed, there being "…40 feet of water in the dam." (Fulton 1880). Therefore his report by itself sheds no light on who pulled the discharge pipes. Their removal must have happened before any new embankment was placed in the center of the dam. And as I noted previously, the 50-man crew began work at the dam while Reilly still owned it.

Pitcairn and other Club members were no doubt wincing at the public assertions that no qualified engineer had been involved in the dam repairs. This of course added energy to possible liability suits against the Club. For example, *Eng. News* (1889a p 550) proclaimed:

… the essential fact will still remain unchallenged and unchallengeable, that no engineer of experience or standing, in responsible charge of work, however humble, lent any assistance in the plans for reconstructing the [South Fork] dam, or was asked to do so, and that no engineer of any kind was permanently employed.

The fact that no engineer at all was consulted as to the dam rather destroys the force of a moral which the Johnstown disaster would otherwise have for engineers,—that the profession as a whole has a deep interest in protecting itself from the disgrace which the incompetence of charlatans and ignoramuses may at any time throw upon it. Supposing that there had been a so-called "engineer" engaged, who could possibly show a sheep skin conferring that title on him, what means would have remained to the profession as a body to disprove the charge which many newspapers have been quick to bring against engineers—that this disaster proves that they are unable to make any such work really safe? There is but one possible means,—that engineers should themselves assume part of the responsibility of deciding who is and who is not fit for "responsible charge of work"; there is but one effective way by which they can certify to this fact, —by maintaining a national engineering society, to which all reputable engineers are expected to belong...

So who really took out the pipes? After reviewing the historical records, Uldis Kaktins and I arrived at the answer in a conversation at the South Fork dam one sunny autumn day. The first falling leaves were drifting with the wind as we stood beside the large shaped stones along the river. This was the foundation of the old control tower, near the upstream end of the former discharge pipes. Uldis looked down at the stones then cast his glance downriver where the pipes once discharged their flow.

> You know Neil, the only answer that makes sense is both Reilly *and* the Club pulled the pipes. It fully explains why a big crew began work at the dam while Reilly still owned the land, before he transferred the deed, and also reconciles the different stories, that either Reilly pulled the pipes or the Club did it.
>
> Yes Uldis, that has to be it! There must have been a simple "handshake" deal between Ruff and Reilly; Reilly would transfer the property but *only after* the Club workers pulled or broke up the cast iron pipes, sent them downhill by wagon to South Fork, and loaded them on flatcars for delivery to the iron mill in Johnstown. After delivery Reilly got payment for the scrap iron and that made up for his selling the property to the Club at a small loss.
>
> Sounds right Neil. Everything they did at the dam needed either horse or man power, and would've been based on common sense. As the land owner Reilly never had to touch the pipes or hire his own crew. That would've been a big expense. And a big crew was needed to excavate the 70 or so tons of iron in those pipes. Ruff had his 50 men on the job five months before the property transfer and a month before the Club's charter was approved in Pittsburgh. In other words, both Reilly and the Club, in effect, pulled the pipes! But the Club's laborers would have done the actual work. As a politician Reilly was shrewd and smart enough to demand this. Ruff wanted the land and the old dam and Reilly made him pay a price in money and labor. Reilly was indeed wily – I recall he was a Republican elected as a Democrat! But they didn't reelect him.
>
> Uldis, this also explains why James Francis and the other investigators wrote that the Club pulled the pipes. They interviewed people living near the dam, some of whom had lived there many years, and those folks would have remembered the 50-man crew of laborers showing up at the dam years earlier and what they did. They probably didn't know Reilly still owned the land then. And if Reilly ever saw Pitcairn's PRR testimony, claiming that the former Congressman pulled the pipes, he wouldn't have disputed the powerful executive from Pittsburgh, especially if removing those pipes was an unwritten part of the deal to transfer the deed. I'm surprised Reilly didn't also ask for a lifetime membership in the Club.
>
> Sure Neil, but Ruff may not have told him about all his plans for the resort.

By March 15, 1880 the Club owned the old dam and the associated land. Work started anew to repair the dam that summer. The *Tribune* (1881 Feb 11) carried a front-page story about "The Old Reservoir." Prophetically, the writer began with the words, "There is a great deal of apprehension manifested by some of our citizens in relation to the stability of the Portage [South Fork] Reservoir..." A group called the "Western Game and Fish Association" had bought the property around the reservoir and repaired the dam. The article went on to say there may be "... some danger should the dam break..." but since the water would distribute itself over the many miles to Johnstown "...there is not much probability its power would be very great here." This story allayed the fears of some people in the downstream towns and boroughs, but certainly not all.

In the newspaper column beside that February 11th story were advertisements from professional men, Doctors J. K. Lee and L. T. Beam, and Justice of the Peace John H. Fisher. All three men would die 8 years later in the flood.

4.3   Influence of Robert Pitcairn, PRR Superintendent, Western Division

The South Fork railway station did not yet exist in 1880, but trains stopped there on occasion because of the local coal mining operations. The South Fork Fishing & Hunting Club officially opened the next year. After the flood, Pitcairn's testimony for the PRR tosses up some confusing historic "chaff." He claims he had reservations about repair of the dam, and that his own engineer, Mr. Halliday, inspected the dam and provided a written report, but Pitcairn was unable to find his copy. It came out later that Robert L. Halliday, who was Superintendent of the Lewiston and Sunbury Division of the PRR, had declared the South Fork dam unsafe (*Eng News* 1889b). Pitcairn lauded "Colonel" Benjamin Ruff, whom he had known for many years. And yet, despite Pitcairn's supposed reservations about repair of the dam, he was one of the earliest members of the Club and visited it within a month of the opening.

Although Pitcairn asserted that he had arranged for "educated" railroad engineers to check on the South Fork Dam, there is no evidence that either he or Ruff had inspections done by engineers experienced in hydraulics and dam construction. Daniel Morrell, General Manager of the Cambria Iron Company, was so concerned about the dam that he had sent his chief engineer Robert Fulton to inspect it. Fulton's letter report to Morrell was dated November 26, 1880. Fulton was both an engineer and a geologist and gave important professional insights about the dam. The content of his letter shows that the embankment repairs were far along by that time, and indeed other accounts refer to the work being finished in the summer of that year. Morrell sent the report on to Ruff, who provided such an uncharitable reply to the letter that he clearly did not take Fulton's comments seriously or did not understand their safety implications. Pitcairn claimed that Morrell "...was with me in sympathy opposed to the rebuilding of the South Fork dam..." Pitcairn also said "The first information that I had on the matter [of the dam] was on the commencement of the rebuilding of the dam..." These two statements confirm that Pitcairn knew of the Club's plans from the beginning and was not learning of this much later after the repairs were made.

After Fulton's visit, one of the objections raised by Ruff was that Fulton's letter gave the wrong name for the Club, calling it the Sportsmen's Association of Western Pennsylvania. But that was the name for the Club as known in Johnstown at the time, and many if not most of the Club members were also members of a club with that name in Pittsburgh. The Club at South Fork was clearly an outgrowth. In fact at the time of his death in 1887, Benjamin Ruff was Vice President of the Sportsmen's Association of Western Pennsylvania. This association shows up in J. F. Diffenbacher's (1895) "Directory" years after the flood, with an address at 605 Smithfield Street, Room 609.

If Pitcairn was truly opposed to the repair of the South Fork Dam, as he devoutly claimed after the flood, he certainly had full authority to bring in consultants who could provide expert opinions on the design and safety of the dam repairs. Although the railroad no longer owned the property, Pitcairn could have justified this for the protection of the extensive PRR property downstream. Even in the 1880's dam engineers would have had major concerns about any plan to lower the embankment, thereby diminishing the spillways. He preferred Ruff's railroad experience (he had also built a prison) over professional expertise. In Pitcairn's judgement, Ruff "… was better than any engineer." But as history would sadly reveal, Ruff was *not* the equal of any engineer and was certainly not better.

4.4   Changes Made to the South Fork Dam

The partial failure of the embankment in 1862 set the stage for the disaster in 1889 because when the dam was rebuilt by the Club changes were made that significantly altered its original design. The changes, such as lowering the dam crest, the omission of low permeability "puddled" clay layers that had originally been emplaced on the upstream half of the embankment, and the possible use of mining wastes containing plastic clays, are discussed by Kaktins et al. (2013).

A critical moment came when Colonel Ruff first acquired the property with the intent of repairing the dam and building a summer resort. Ruff himself was proficient in building railroad embankments, some of which can be quite large, but he was not an engineer and had no experience I can discover that related to large or even small dams. After acquiring the old dam and the surrounding property he needed to consult with professionals with expertise in the construction and repair of large dams. The partial breach from 1862 revealed the internal structure of the original embankment, with clear evidence for an engineer of the construction methods used and the quality of the results. A proper dam engineer would also have advised on the repair plan, the adequacy of the spillways, and whether discharge pipes at the base of the dam should have been retained or replaced if removed. An engineer with know-how in hydraulics would have strongly counseled against lowering the crest of the dam because of the resulting big reduction in spillway discharge. But there is no evidence I can find that Ruff or anyone else brought in a qualified and experienced engineer to consult on the repair of one of the larger dams in the world. Because of this unfortunate but intentional oversight, Col. Benjamin Ruff bears a historic share of blame for the 1889 dam breach and catastrophic flood. But he could not be held legally liable because he had died at the Monongahela House in Pittsburgh, 2 years before the 1889 flood (*Post* 1887).

As previously discussed, removing the discharge pipes at the base of the dam was a significant change. The upper part of the stone culvert where the pipes had emptied had collapsed in 1862, but a downstream section of the culvert remained along with part of the overlying embankment. The workers blocked the opening of the remaining length of culvert with vertical hemlock planks and proceeded to build the embankment. Pulling the pipes meant that the Club could no longer control the water level in the lake, and reduced the overall discharge capacity. Permanent removal of the pipes was bad enough, but another change – lowering the dam crest – would truly doom the dam on the day of the flood.

The top of the dam as originally built was only 10 feet wide, adequate for riders on horseback to cross but more difficult for a wheeled horse-drawn cart, wagon, or carriage. Certainly wheeled vehicles could not pass each other on the narrow crest road. The story comes down to us that the crest was lowered to widen the carriage road (Francis et al. 1891 p 446). Lowering the crest certainly achieved that, but was *not necessary* to easily widen the road. Additional stone and fill material with cribbing could simply have been added to the uppermost slopes of the dam, widening the road by the amount desired. Benjamin Ruff would probably have seen that approach as a waste of material that could go into dam repair. In reality, digging and scraping tons of material off the crest was an obvious way to quickly get fill material to start repair of the partial breach of 1862 (Coleman et al. 2016). From Ruff's viewpoint, lowering the tops of the dam remnants also meant that much less material was needed to fill the breach up to the new reduced height. Uldis Kaktins and I realized these were the primary reasons the top of the dam was lowered. Had Ruff consulted with any engineer experienced in dams he would have been told lowering the crest would endanger the dam, and why. Lowering the crest was a *fatal* mistake.

The original builders of the South Fork Dam had the five large drainage pipes to pass the stream flow while they built up the embankment. Even so, major storms could produce runoff that easily exceeded the pipe flow capacity so they also used planking structures. Using the methodology of Chow (1964) for pipe flow, I estimate the five pipes together could have conducted, under low head conditions (~1 m), about 154 cfs (4.4 m^3 s^{-1}). Low flows on the South Fork today can be as low as 20 cfs, but even moderate rain events can rapidly spike the discharge to >150 cfs (personal communication, C. Coughenour, 2017, Univ. of Pittsburgh at Johnstown).

The problem for the Club's laborers was that there were no longer any pipes to carry the stream flow so the lake would continue to rise. While the embankment was being raised all the flow had to be shunted through a series of temporary spillways made of planking. I have not discovered when the dam crest was lowered, but Ruff wanted to proceed quickly so it would likely have been done to support the initial repairs in 1879. If so, what happened that Christmas week of 1879 was most unfortunate, when high stream flows washed out the repairs made to date (*Tribune* 1879b). Any "benefit" in time or cost gained from scraping and dumping the top of the dam remnants into the breach was gone. The lowered dam was now destined to fail and the towns below it were doomed.

Ruff was forced to halt work until the next year, 1880, when considerable progress was made in filling the partial breach of 1862. Ruff wrote that 22,000 [cubic]

Fig. 4.2 View along crest of the eastern dam remnant, looking westward toward the observation platform. This part of embankment represents original construction but with lowered crest. Large riprap are still in place, some standing more than 0.6 m higher than present crest due to embankment being lowered as much as 0.9 m by the Club. White stick on rocks is 1 m long. Photos by the author, May 2017

yards of material were needed to fill the breach. John Fulton visited and inspected the dam that November. Ruff planned additional work for the spring of 1881, perhaps partly in response to Fulton's review, but the Club first had to repair a small break in the dam caused by a February flood (Unrau 1980 p 72).

The lowering of the dam crest is nicely shown even today by the fact that large riprap boulders on the intact eastern embankment stand above the present crest (Fig. 4.2). The riprap boulders at the highest part of the dam face would have been placed with their tops equal to the dam crest elevation, not higher than the crest as they appear in the figure. Therefore the present large stones illustrate the degree of crest lowering on the surviving parts of the embankment. Only a few large riprap remain at the crest on the western dam remnant. The western abutment was accessible by road many years after the flood and according to locals served as a "lover's lane." Vehicles could drive onto the top of the dam there using part of the old carriage road. Some of the riprap along that western crest were long ago tumbled down the bank or removed for other purposes. Prior to the 1970's a road also passed through the eastern spillway, but cars and trucks could not easily reach the eastern top of the dam.

Additional changes made by the Club included construction of a bridge over the mouth of the spillway and the installation of a boom and fish screens at the bridge. These features could have had the effect of somewhat reducing the main spillway discharge capacity if the screens became severely clogged with debris during a flood. However, the narrowest part of the spillway was about 100 ft. north of the foot bridge and likely controlled the cross section of flow.

Another change relates to the masonry stones that were originally placed on the upstream side ("slope wall") of the dam. As shown in Fig. 3 of Kaktins et al. (2013), the placement of these dressed stones on the northeastern side of the dam seemed to be incomplete at the time of dam failure. According to Benjamin Ruff, the face of the dam on the lake side was not covered with riprap, but was faced with a "slope

Fig. 4.3 Springhouse at the Unger House. Photo reveals that dressed stones were used in its foundation, likely scavenged from the upstream slope wall of the South Fork Dam. Photo by the author in 2017

wall" (McGough 2002 p 24 and *Tribune*, 1889). Therefore the masonry cover that comprised the slope wall may well have been complete for the original dam, but some of those dressed stones were apparently removed during or before the Club's reconstruction. A National Park Service (NPS) image reproduced in McGough (2005, Illustration #9) shows that, at least on the lake side of the northeast remnant, the original dressed stones had been replaced by riprap on the rebuilt dam.

Masonry stones that appear similar in size to those in the dam's slope wall exist in the foundation of the house and springhouse of the former Unger property, on the north side of the dam (Fig. 4.3). Unger was the last president and manager of the Club. The former barn reportedly also had dressed stones in its foundation. They could easily have been scavenged from the dam before it was rebuilt. Photos of the lake side of the dam after the flood show that the remaining original slope wall covered less than the bottom fourth of the slope on the northeast remnant of the dam. It would have been very unusual in the 1800's to fabricate dressed stones to use in the foundations of a barn or springhouse, unless such stones were already available nearby. It should be noted that the springhouse foundation shown in Fig. 4.3 is not entirely original. Parts were restored by the NPS. The present foundation is likely a good representation of the original, given the NPS's goal to accurately preserve historic sites like this.

Another important change by the Club relates to the large riprap originally placed on the downstream slope of the dam. Riprap forms a protective layer to prevent erosion and ensure stability of the embankment. The Club *did not* place large riprap like those in Fig. 4.4 to cover the repaired section. Smaller riprap was used, and this is evident in Plate LIIIA of Francis et al. (1891) (see Fig. 4 of Coleman et al. 2016, available online), which is an undated image of the dam after the Club repaired it

Fig. 4.4 The author standing on original riprap (Conemaugh Group sandstone) preserved on downstream side of the dam remnants, placed in late 1851 or 1852. Western dam remnant is at left and eastern remnant at right. By contract the riprap layer had to be at least 4 feet thick at the top of the dam and 20 feet thick at the base of the slope. Each placed stone had to equal or exceed a volume of 4 cubic feet. Photos by the late Uldis Kaktins

but before the 1889 flood. That plate was included in the discussion section of Francis et al. (1891) which means it was provided to the investigators after they wrote the ASCE report. Nonetheless, Francis et al. (1891 p 446) did not modify the judgement in their write-up that "heavy" riprap was used to cover both sides of the repaired embankment. But it simply was not true.

The photograph of the pre-flood dam (Coleman et al. 2016, Fig. 4) had been submitted by engineer Francis Collingwood for consideration by the ASCE investigators. When their report was published in 1891 Collingwood was secretary of ASCE and would hold that title until 1894. By this time he was known as an eminent professional, having served as an assistant engineer to Washington Roebling during construction of the famed Brooklyn Bridge, completed in 1883.

The original South Fork Dam, instead of a "heart" wall of masonry, relied on "puddled" clay layers to ensure low permeability integrity of the embankment. But the puddled layers were not replaced during the Club's repair work. The embankment was built up using random fill materials, basically dumped into the breach. Francis et al. (1891, p. 445,446) gave additional details about the Club's work. My notes are shown in brackets:

> In April 1880, the work of repairing the breach in the dam of July, 1862, was commenced by the Hunting and Fishing Club [*not so - the initial work began 6 months earlier in mid-October, 1879 with a 50-man crew*]. It was let out by contract to Colonel B. F. Ruff. The five lines of 24-inch sluice pipes were taken out [*this is supporting evidence that the Club*

workers removed the pipes], the masonry on which they were laid and the remains of the culvert were left. A sheet piling of plank was put across the lower part of the breach. The original plan of making the lower angle of the embankment of stone was adopted, and stone of as large size as could be obtained in the vicinity were dumped into the breach, letting them form natural slopes. This stone embankment was carried up until it reached such a height as to enable a road to be graded down from the crest of the remaining parts of the dam on each side of the breach, so that material could be hauled in carts from the borrow-pits on the hill side. There being no sluices for the discharge of the flow of the stream, the surplus water found its way through the stone embankment, the water in the reservoir rising as the filling on the upper side of the stone embankment proceeded. The washing of the filling through the stone embankment was prevented by covering its face with brush, hay, etc. [*this organic material would not have prevented laterally infiltrating water from transporting clay and silt through the stone layers*]. The material relied on to form the water-tight embankment consisted of clay and shale, which was dumped in on the upper side of the stone embankment, and carried up in layers to the full width of the remaining parts of the original dam. There was no systematic puddling done, but the hauling by teams over the freshly deposited material, which was kept wet by the rising water, made a fairly compact embankment on the on the upper side of the stone embankment [*this was no substitute for proper "puddling" work*]. The work was not completed that season, and during the following winter it was damaged by a flood, and the next year, 1881, the Hunting and Fishing Club completed it by day work. The slopes on both sides of the embankment were covered with a heavy rip-rap [*not correct, as shown in Plate LIIIA of the Francis et al. report this quoted text comes from*].

4.5   Opening of the South Fork Fishing & Hunting Club

The Club finally completed repairs on the South Fork dam and the lake gradually rose until it reached full depth. The Clubhouse and other buildings were nearly ready - by July 1881 the Club officially opened for business. Portions of the guest register from the check-in desk are preserved in the archive of the Johnstown Area Heritage Association (JAHA). The register tells us who the first guests were and gives a sense of who frequented the Club during its early years. Some prominent names of Club members never appear, such as Andrew Carnegie. But Carnegie's name may only be missing because some 73 pages were long ago cut out of the bound book. They have not been recovered to date.

Some member names appear regularly in the ledger such as Benjamin Ruff, Christopher C. Hussey and family members, and the family of Henry Holdship, but others do not appear at all except in the list of 60 members in the back, such as Carnegie, or may have been in the pages of missing ledger. Pitcairn shows up on page 1 and also visited on June 29, 1883. It may be that, over the years, members were not always expected to sign the ledger on arrival, and would not do so if visiting the private cottages. Final preserved entries before pages were cut out were from June 1886. There is also a gap with no removed pages from September 1883 to the spring of 1886. A different ledger might have been used during that period. In 1886 the only entries were for men boarding and lodging at the clubhouse.

The first guest to register on page one of the clubhouse ledger was John D. Hunt of Pittsburgh, on July 28, 1881. Robert Pitcairn's name also appears on page one when he arrived on August 25th in the company of fellow Pittsburghers and members Henry Frick, John J. Lawrence, Edwin Myers, and John Hunt. Hunt was returning after apparently enjoying his visit a few weeks earlier, although that visit might also have served to "check out" the accommodations for the group. The South Fork train station did not yet exist, but Pitcairn as railroad superintendent would have directed the stop at South Fork for his Pittsburgh group. The South Fork station was built 2 years later in 1883, a certain convenience for the Club. That station would not have been added as a passenger stop without Pitcairn's orders. He undoubtedly used his influence to have the South Fork station built just a few miles from his Club's property, an easy carriage ride away.

"Colonel" Unger and Benjamin Ruff both checked in on August 17th and might still have been at the Club when Pitcairn arrived. Unger did not yet own the "Lake View" house east of the dam, which he bought from Joseph Leckey in December, 1885 [ostensibly for the Club but he chose to keep it; this "Unger house" is today near the NPS visitor center; Leckey had been a caretaker of the dam for the PRR].

The Club operated successfully for 8 years. At some point a telephone line was run from the clubhouse down to the dam, then along the crest road and down to South Fork, but it was only operated during the summer season. The lake served its purpose for many wealthy families as an idyllic escape from the summer heat and clouds of soot and smoke in Pittsburgh. The people and their children and friends fished and swam in the lake, sailed and rowed boats, rode on a small steamer, or simply walked along charming lakeside paths. Young people met the children of other wealthy families, and businessmen no doubt found the bar and porch of the Club to be relaxing places to enjoy pipes, cigars, and scotch whiskey, and to discuss future ventures. Visiting the Club grounds also brought families safely away from the city where outbreaks of typhoid fever and other maladies were all too common, mainly from contaminated water supplies.

There exists in the JAHA archive a printed example of the lake activities planned for the young people on Saturday, August 22, 1885. This was probably in the way of a special "end of summer" celebration, and an example of festive times at the Club. Titled "Regatta and Feast of Lanterns," the afternoon program described boat races for canoes, single and double sculls, including single sculls for girls and ladies, and to generate mirth, a "tub" race for 10 boys over a distance of 60 yards. That evening a "Feast of Lanterns" began at 8:00 with a procession of 50 boats towed around the lake by a small steam launch. Each boat was lit up by 5 lanterns, which would have made a memorable sight. During the towed procession fireworks were set off from the center of the lake. Afterwards the young people landed their boats and gathered in the clubhouse parlor to receive their regatta sporting prizes.

In May of 1889 the Club was preparing for its 9th season on the lake. We now know that less than a week before the 1889 flood the Club took out a mortgage of $34,000. That and a bond sale were intended to finance some Club improvements. Richard Burkert, Director of the JAHA, suggested that the monetary instruments may have been sought to finance the ongoing plumbing and sewer work (Hurst 2016), although additional Club infrastructure may also have needed funding.

Fig. 4.5 Clubhouse of the South Fork Fishing & Hunting Club in May, 2017, at St. Michael, PA. Photo by the author

The improvements were indeed necessary, and extensive. Up until that time the sewer facilities were probably on-lot drainage systems, feeding into vaults with drains that had begun to fail due to excess volume and the local dense, clay soils. There may have been unsavory surface outflows as greater numbers of people visited during peak summer seasons. There would also have been some privies located on the Club property at points of convenience for guests and for use by laborers. The Clubhouse (Fig. 4.5) had 47 bedrooms and there were at least 16 cottages, one occupied by four families (Unrau 1980 Appx. Q). An article by the Tribune (1885) noted that the number of people visiting, including members, their relatives, and friends, was growing yearly. The Clubhouse prepared meals for all the guests, but "…the seating capacity of the dining room is only ninety, however, so that the guests are greatly inconvenienced and the Club House people nearly driven crazy at each recurring meal time." To further add to the seasonal throng, three new cottages were erected in the summer of 1888. At least 200 people were expected for the 1889 season.

According to Unrau (1980), the plan to solve the drainage problems was to trench and install a wrought iron sewer line from the Clubhouse that would run nearly a mile along the lake shore to beyond the dam. Service lines from the cottages would link to this main line. The work began in April 1889. W. Y. Boyer, the Club's Superintendent of the Lake and Grounds, commented that the new sewer work was nearly complete by the time of the dam breach, terminating at a "waste works." He estimated the cost of work to date at $12,000 (Unrau 1980 Appx. Q).

A young engineer from Philadelphia named John Grubb Parke, Jr. had been hired to survey and supervise the plumbing and ditching for the sewer lines and to oversee the installation and initial operation of the drainage and waste facility. Parke had taken course work in civil engineering at the University of Pennsylvania but did not complete his degree before leaving to work with the railroad. Several years later he got the job with the Club. The sewer project was nearly complete by the last week of May, a necessity for the Club to be ready for the many expected guests. Thursday, May 30, was a national holiday, formerly called Decoration Day. Unless they were behind schedule, John Parke and his crew of laborers would have had that day off from their engineering project. But Parke could not have known that in a matter of hours he and others at the Club would witness one of the great tragedies of the nineteenth century.

The breach of the South Fork Dam would kill as many men, women, and children as the reported combat deaths for Union and Confederate troops at the Battle of Fredericksburg, 1862.[4]

References

Chow VT (ed) (1964) Handbook of applied hydrology. McGraw-Hill Book Co., NY, 1418 p

Coleman NM, Kaktins U, Wojno S (2016) Dam-breach hydrology of the Johnstown flood of 1889 – challenging the findings of the 1891 investigation report. Heliyon 2 (2016), 54 p, https://doi.org/10.1016/j.heliyon.2016.e00120, e00120

Connelly F, Jenks GC (1889) Official history of the Johnstown flood, illustrated. Journalist Publishing Co., Pittsburgh, 252 p

Coughenour C (2017) Personal communication, preliminary measurements of discharge on the South Fork of the Little Conemaugh, Sep. 28 and Nov. 6, 2017. Department of Energy & Earth Resources, University of Pittsburgh at Johnstown

Diffenbacher JF (1895) J. F. Diffenbacher's "Directory of Pittsburg and Allegheny Cities" Stevenson & Foster Co., Printers, 527 and 529 Wood St, Pittsburg PA

Eng. News (1889a) Engineering News, Jun 15 1889 p 550

Eng. News (1889b) Engineering News, Jun 8 1889 p 518

Francis, JB, Worthen, WE, Becker, MJ, Fteley, A (1891) Report of the committee on the cause of the failure of the South Fork dam. ASCE Trans. Vol XXIV, 477, p 431–469

Fulton R (1880) Letter report to D.J. Morrell concerning South Fork dam, dated Nov. 26 1880

Hurst D (2016, June) 1889 Flood's mystery: why did club members take out mortgage days before disaster? *Tribune Democrat* 19 2016

Jordan JW (1921) John Reilly – Man of large affairs. In: Encyclopedia of Pennsylvania Biography (Illustrated with portraits). Lewis Historical Publishing Company, NY p 82–84

Kaktins U, Davis Todd C, Wojno S, Coleman NM (2013) Revisiting the timing and events leading to and causing the Johnstown flood of 1889. PA Hist J Mid-Atlantic Stud 80(3):335–363

Kooser NM (2013) Determining responsibility for the failure of the South Fork Dam, unpublished thesis, American Public University System, March 11 2013, Charles Town WV

---

[4] As with other large engagements of the Civil War, many of the thousands wounded would die in the following weeks and months due to the primitive state of battlefield medical and surgical care.

LDA Architects and Wallace, Roberts, & Todd (1993) Historic Structures Report, South Fork
    Fishing & Hunting Club, St. Michael, PA. Appendix A.3. (Property Trans.). Purchases by
    South Fork Fishing & Hunting Club, 1880–1887. Prepared under contract to The National Park
    Service, Denver Service Center for The Southwestern PA Heritage Preservation Comm. and
    The 1889 South Fork Fishing & Hunting Club Historical Preservation Society
McGough MR (2002) The 1889 flood in Johnstown, Pennsylvania. Thomas Publications,
    Gettysburg, PA, 180 p
McGough MR (2005) The Club and the 1889 flood in Johnstown, PA. Friends of the Johnstown
    flood National Memorial, 124 p, illustration #9
Post (1887) Pittsburgh Daily Post, Edition of Thursday, March 31, 1887
PRR (Pennsylvania Railroad) (1889) Testimony Taken by the PRR Following the Johnstown Flood
    of 1889. [Statements of PRR employees and others in reference to the disaster to the pas-
    senger trains at Johnstown, taken by John H. Hampton, at his office in Pittsburgh, by request
    of Superintendent Robert Pitcairn; beginning July 15th, 1889.] Copy in archive of Johnstown
    Area Heritage Association. Many stories available online at: https://www.nps.gov/jofl/learn/
    historyculture/stories.htm. Accessed 4 Jan 2018
Tribune (1879a) Johnstown daily tribune, Wednesday, Oct 15 1879
Tribune (1879b) *Johnstown Daily Tribune*, Dec 27 1879. Note about the old reservoir and Western
    Game and Fish Assoc. under "Local Items" Vol. VII, No 258, p 4
Tribune (1881) *Johnstown Daily Tribune*, The old reservoir. Editor G.T. Swank, Friday Feb 11,
    Vol. VIII, No. 296, Johnstown, PA
Tribune (1885) Johnstown Daily Tribune, Monday, Aug 10 1885
Tribune (1889) Johnstown Daily Tribune, Tuesday Jun 18, 1889) Vol XVII No. 5081. Reprint of
    letter dated Dec 2, 1880 from Benjamin F. Ruff to Daniel J. Morrell
Unrau HD (1980) Historic structure report: the South Fork dam historical data, Johnstown Flood
    National Memorial, PA. US Department of the interior, NPS, package no. 124, 242 p

Chapter 5
Day of the Dam Breach

5.1   The Rail Journeys of Michael Trump and Robert Pitcairn

Thursday, May 30th, the day before the flood, was a holiday previously called Decoration Day. By 1889 May 30 was the national holiday called Memorial Day. Robert Pitcairn, Superintendent of the PRR's western division, was at work in Pittsburgh but went home early, about two o'clock. He had a telegraph instrument in his home and was a skilled operator himself, going back to his early days with Andrew Carnegie. He made inquiries through the lines until about 11 o'clock, being concerned about reports of heavy rain. Heavy rain usually meant washouts and train delays.

The next day started badly at the Pennsylvania Railroad offices in Pittsburgh. The telegraph lines were clicking messages about track problems east of Johnstown, beyond South Fork, high on the plateau near the town of Lilly. Pitcairn came to work on the early train and found the Chief Dispatcher, Charles Culp, waiting with reports of flooding at Lilly. The dispatchers at Union Depot had nearly direct control of all the movements of trains on the division, under Pitcairn's name. Pitcairn spoke with Michael Trump, his Assistant Superintendent,[1] and decided to send him quickly to the east with any men he needed to make emergency repairs, especially to fix downed telegraph lines.

Soon more alarming news came into Union Depot from the AO telegraph tower, east of Johnstown and beyond East Conemaugh, just west of a large loop in the river crossed by Bridge #6. The telegraph lines between AO and Mineral Point were down, but a new message had been hand carried from Mineral Point down to AO, then relayed to Pittsburgh. It was the first message about the South Fork Dam and this dispatch was given personally to Pitcairn by Charles Culp. In his PRR testimony

---

[1] Michael Trump would become PRR's General Superintendent of Transportation in 1897.

© Springer International Publishing AG, part of Springer Nature 2019
N. M. Coleman, *Johnstown's Flood of 1889*,
https://doi.org/10.1007/978-3-319-95216-1_5

Culp was not sure how many messages about the dam came in to the Depot.[2] He was asked what information, if any, he got about the South Fork Dam and to whom it was sent:

> Well, now, I am under the impression that we received two messages. I think there was one received about 12 o'clock [noon]… I think it was directed to the Superintendent [Pitcairn].

When asked about the contents of the messages and the time received he replied:

> I don't remember the contents, other than it referred to the probability of the South Fork dam bursting…. I think that was about 12 o'clock, or shortly after. …. I took the first message over, and laid it on Mr. Pitcairn's table in front of him, and I think we received another message after two o'clock after Mr. Pitcairn had gone, to inform the people at Johnstown that there was a probability of the dam bursting. That message was sent to Mr. Pitcairn on his special car. I wouldn't be positive about it, but I think there were two messages received.

Charles Culp reported that Trump left at 11 o'clock on his train, and Mr. Pitcairn went in his special car on the rear end of a regular train. Trump had messages in hand about tracks washed out far to the east at Lilly, between Portage and Cresson, and was worried the telegraph lines might go down which would cut communications to the east. He was right to worry - it was common for washouts to push down telegraph poles. That very thing happened that morning east of Bridge #6, as reported by Mr. William Reichard, causing the line break between the AO tower and Mineral Point.

The telegraph operator at AO tower, Mr. Shade, gave testimony as to the content of the first message, the one that was laid on the desk before Pitcairn. As he recalled, it read "The waters at South Fork dam are very high and it is liable to break at any moment; notify people of Johnstown and vicinity to prepare for the worst." By Pitcairn's own testimony, he knew this message came from Unger, President of the South Fork Fishing and Hunting Club.

Michael Trump – On Train from Pittsburgh to Johnstown Trump moved fast with an engine and one coach. With him were several assistants, a master carpenter, telegraph operator, and a gang of telegraph repairmen. They eventually reached Johnstown and saw high water in the streets. They stopped to check the abutments at the stone bridge[3] below the town and then went on to the telegraph tower at the railway station. Trump later recalled that the Freight Agent at Johnstown, Mr. Frank S. Deckert, had:

> …informed us that he had a report the dam was liable to break at any minute, and he said the people of Johnstown did not seem to realize the thing, didn't believe it, and paid no attention to it; that they had heard it so often that they placed no confidence in the report. I then advised him to get his family out of the house, and try to persuade everybody else to get out that he could.

---

[2] At least three messages about the dam were eventually sent.

[3] This was the Pennsylvania Railroad stone bridge below Johnstown. This 7-arch bridge had been built about a year before the flood to permanently replace an iron bridge.

If I may just say right here, a precious chance to save lives was lost. If Michael Trump felt that Freight Agent Deckert and his family were in danger, on ground higher than much of Johnstown and far from the dam, he should at least have warned railroad staff and passengers up the line, closer to the dam, especially at East Conemaugh where he knew passenger trains had been held along the river. And what of the local people close to the rail line in East Conemaugh, Franklin, and Woodvale? The telegraph lines were open from Johnstown to beyond East Conemaugh. Lots of Trump's men worked at the yard and engine roundhouse. Nearly 30 locomotives and other equipment were there, along with three passenger trains held up by the track damage to the east. And why did Trump lay the responsibility on Deckert to warn Johnstown when Trump himself was a senior railroad executive? Trump had been told that local people were not heeding the warnings. His name and position were well known in Johnstown, and messages or verbal warnings directly from him or by his authority under Pitcairn's name would have been taken far more seriously. Of course, much depends on the authority granted by Pitcairn to his Assistant Superintendent. How tight a leash did he keep on Michael Trump?

Frank Deckert – Johnstown Rail Station For several hours before the flood wave arrived, Deckert stayed at his post receiving and sending messages and notifying people near the station. He sent a warning about the dam to the Central Telephone Exchange in Johnstown, run by Mrs. Hettie Ogle. She also managed the Western Union office. We will never know how many people got and heeded the warnings and left the low areas. Some lives may have been saved. By then flooding in the lowest streets was already high enough that many would have had to wade deep water to escape - in some places needing a boat. Most of the people who left had done so earlier in the day. Then Deckert saw some people raise a commotion on the station platform and someone yelled "fight." He then

> …went right down to my house as hard as I could run, though I didn't really know what was the matter, -- and why I ran, I don't know, but I went to the house, and -- went in and rushed them [my family] out just as I came to them, and then I came out myself last, and I no more than got outside the gate when the houses were falling around me on the other side of me, but they apparently fell without any cause as I could see, and I didn't understand it at all…

Michael Trump – East Conemaugh From his testimony to the PRR, Deckert got his family out in the nick of time but he does not mention Trump's visit to his station. Trump had gone, directing his train on to East Conemaugh where he could seek more information upriver. He gave the following testimony:

> On arrival at the telegraph office at [East] Conemaugh, we received a further report from the Yard Master that the condition of South Fork dam was getting serious, and the dam was liable to break at any minute. I looked over the ground with Mr. Walkinshaw, and found that he had already placed the passenger trains, first and second No. 8's, on the back tracks in the yard, and in the safest position in which they could be put. We then received a report that the tracks were washed out at the water works dam east of Bridge 6, and we concluded the best thing we could do was to push on and see what the trouble was there especially the telegraph line east of Conemaugh was down, and we hadn't the men there to repair it.

Trump's last comment is very odd because he brought a whole "gang" of telegraph repairmen with him, and from available messages he would have known the line trouble was between his present location and Mineral Point. He may have become less alarmed after seeing the Little Conemaugh River had not yet overflowed its banks at East Conemaugh. But around half past 3 o'clock he had personally seen the dispatch that the South Fork Dam was liable to break at any minute.

As yet unknown to Trump the dam had already breached. He believed the Day Express trains were "in the safest position in which they could be put." Sadly, another chance to save lives was now lost. Passengers are not freight – they can move themselves. Although it was raining off and on, if the passengers had been informed that an upstream dam was "…liable to break at any minute…" they at least could have chosen to stay in the train and risk the dam break or move to higher ground and get wet in the process. And no doubt the heavy rains had created a sea of mud beyond the railway platform and boardwalks. Then Trump's engine and car moved on east past Buttermilk Falls[4] to approach the AO tower. The flood wave was on its way.

Three trains with passengers were in the path of that flood wave. The mail train and two passenger trains of the Day Express had been delayed and were sitting idle at East Conemaugh, just two miles upriver from Johnstown. The mail train was on track 1 beside the river. Section 1 of the Day Express was several tracks nearer the hillside. Section 2 was closest to the hill on an embankment 5 to 6 feet high, although there was a wide ditch between that train and higher ground. When the flood wave struck it destroyed the large locomotive roundhouse and smashed Sect. 1 of the Day Express. Everyone on the mail train had been warned by their conductor about the possible dam break, more than an hour before the flood wave hit. They were ready to go and all survived, despite having the farthest distance to run. The mail train was pushed by the flood wave and did survive intact, but its passengers and crew could not have anticipated that. Those who stayed inside Sect. 2 got very wet but miraculously survived. They were lucky. Those who remained on the other Express perished, as did some who ran for the hillside. But sprinting was the only chance for those inside Sect. 1, as the wave tore their train apart.

Charles Warthen, Conductor – On Mail Train, East Conemaugh Most of the rail passengers at East Conemaugh were not explicitly told about the possible dam break, although there were rumors among some of them and the railroad men. But Conductor Warthen did tell his passengers on the mail train, most of whom were a company of actors heading to Altoona. He told them about the warning message that came to the tower, that the South Fork dam "…was liable to break at any moment…" Here is part of his PRR testimony:

> *Q. During the time you lay there, did you hear anything said by Mr. Walkinshaw or anybody else as to the danger of the dam giving way?*
> A. Yes, sir, there was a message came. The operator told us that he had a message from the dam saying that it was liable to break at any moment, and we went out then in the train,

---

[4]The sweeping curve east of Buttermilk Falls was known among the railroad men as the "Buttermilk Curve."

and talked with the passengers about it, but I had never seen the dam, and didn't know how bad it would be if it did break. I told my passengers, in case there would be any trouble of that kind, to be ready to get out of the train. I didn't know what the danger might be. The only warning we had of the dam breaking, and the water coming down, was an engine whistling up the track. It commenced to whistle an alarm, and we looked out and saw the water coming, and hollowed to the passengers, and we just had time to get them out and get on the hill.

> Q. *You had a theatrical troupe, didn't you?*
> A. Yes, sir; "Night Off" Company.
> Q. *Had you any other passengers in the train?*
> A. Yes, sir; they all escaped.
> Q. *What time was it the operator told you he had that message about the dam?*
> A. That was along about three o'clock.
> Q. *How long was it until the flood came?*
> A. I think it was along about 3:30 somewheres.
> Q. *Did you hear of any more than one message?*
> A. No, sir, just the one.

More passengers on the Day Express trains would surely have survived had they gotten the same warning as folks on the mail train. They would have had more time to run. Conductor Warthen's passengers escaped because he had warned them and they were ready to dash, baggage in hand. When the whistle warning came from upriver they got off the train and ran for the hill. They had the farthest to run, but made the best of the warnings and lived.

Robert Pitcairn's Journey – Train Departing Pittsburgh   In his PRR testimony, Pitcairn recalled leaving Pittsburgh around 1 p.m., his special car being attached to the #18 train. He intended to go as far as Lilly. Along the way he was handed dispatches that the trouble with "the water" was getting more serious. At this point in his testimony Pitcairn makes an extraordinary statement. "… *I understood that morning before I started* that the people at Johnstown were warned out by Mr. Unger…".

Pitcairn had read a telegraphic warning about the dam that he knew was from Unger, the President of the South Fork Fishing & Hunting Club to which Pitcairn belonged. But in his PRR testimony Pitcairn went on to suggest he doubted the message because Unger did not have a communication line from the Club to South Fork, and there had been exaggerated concerns about the dam every year. But Pitcairn also knew that Unger lived on the hillside right above the dam - he was in the best place to know its condition.

For three critical hours Pitcairn chose not to send a specific warning *under his name and title* to the people of Johnstown. It is true that in storms over the years the people there had been cautioned many times about the dam, and after so long became inured to them. But they had *never* been warned by Pitcairn himself. His was a household name - nearly everyone knew his name and title. A warning from the well-known Superintendent of the Western Division would have carried great influence and would certainly have spread the word faster and led many more to try to evacuate.

Operator Pickerell at Mineral Point had gotten another message through to East Conemaugh by 2:35 pm, via wire this time, which would have reached Pitcairn by 3:00 from towers along his route, even on his train. This dire message was sent to Pitcairn from his friend at the Argyle mines. It said "The dam is becoming dangerous and may possibly go. J. P. Wilson." Here was the last chance for Pitcairn to send a clear warning under his name and title to everyone in reach of the telegraph lines. But he hesitated, even though he now had in hand forewarnings from both Unger and Joe Wilson, the South Fork mine superintendent he most trusted to keep an eye on the dam. There was still time for Pitcairn to flash a message to East Conemaugh. But it was almost too late – the flood was on its way. In another hour the heart of Johnstown and neighboring boroughs would be wiped out.

Even in the worst case he might have imagined, Pitcairn could not have envisioned the destruction that would befall Johnstown. He must have thought the distance from the dam to the town was so great that any flood would be spread out and lessened. Pitcairn claimed he had had concerns for years about the dam and asked various people to keep him informed about it. He already knew of severe damage at Lilly that day and the loss of communication due to flooding beyond East Conemaugh. He was normally a decisive executive, making many decisions each day. He was also a micromanager,[5] infused with all the big and little details of his railroad division. But Pitcairn's hesitation on May 31, 1889 cost lives. At the very least, he bears responsibility for the lives of railroad staff, passengers, and many town folk lost in the river valley above Johnstown. Guilt about this disaster surely followed him for the rest of his life. It also accounts for extraordinary actions he took after the flood to rapidly repair the rails and bring aid to survivors.

Pitcairn's train moved on until it reached Bolivar where he saw the river at high flood stage. For some reason he attributed this flood to high water in the Stony Creek and the North Fork of the Little Conemaugh. So he said in his testimony, which is odd considering the South Fork dam and its basin are sandwiched between the two.

His train stopped four miles from Johnstown at the Sang Hollow telegraph tower. All signals had been lost to the east and his train therefore had no clearance to go farther. Pitcairn was deciding whether to override the safety procedure when he saw lots of debris coming downriver in small pieces. The flood level rose and telegraph poles were breaking down. The tower itself would soon be threatened. The next thing he saw was a man perched on debris coming downriver. Some men tried to rescue him but failed. Soon after that scores of people were spotted coming downriver, including women and children. The tower had to be abandoned as the river kept rising. Pitcairn's men were only able to save seven persons out of more than a hundred that rushed by. These few said they were from Morrellville and Cambria City, which for some reason made him believe "...the trouble was with the Stony

---

[5]There is a newspaper account *(Tribune,* Oct. 1, 1879) from Johnstown about rail ticket prices. PRR agent Nichols telegraphed Superintendent Pitcairn about more requests for reduced rate excursion tickets to attend a Pittsburgh Exposition. Pitcairn replied in minutes telling Nichols to sell no more tickets at the special price.

Creek water..." But that made no sense because both villages were below the junction of the Little Conemaugh and Stony Creek Rivers in Johnstown. Either or both rivers were to blame. Pitcairn's mind had shut out the awful thought that the South Fork Dam was responsible.

Pitcairn went back to the train and was prepared to have the passengers ready to run for the hillside, but then he noticed the river had stopped rising and slowly started to drop. He then held the freight trains at Sang Hollow and had his own train take the passengers west to New Florence. Those who could get lodging were welcome to stay there and the rest were taken on to Pittsburgh. From New Florence he wrote to the editors of many morning papers to call a public meeting in Pittsburgh because he had "...seen enough people go down the river to know that there would be terrible distress." Later that night he got the terrible news from Mr. Hays, in charge of the work train below Johnstown, "... notifying me that Johnstown was swept away." Pitcairn immediately sent an order to "...collect all the men from the western lines, and material to repair the damage."

After reaching Pittsburgh on Saturday, Pitcairn contacted the General Manager and President of the Pennsylvania Railroad to report the facts and his call for a public meeting. The PRR donated $5000 for victim relief and put all available equipment and men to work to reestablish the rail line and bring help to the Conemaugh Valley. Pitcairn had earned a fine reputation years earlier, during the Civil War, when he had the difficult job of coordinating rail movements of men, supplies, war material, and livestock through western Pennsylvania. His decisiveness now suddenly appeared and worked wonders. So rapid was the response of the railroad that in 5 days a line to Johnstown was repaired and Clara Barton arrived on Wednesday with many supplies and medical staff. She stayed for 5 months (Barton 1910). But even more extraordinary, the line from Pittsburgh to Altoona was open by June 13th. Huge wooden trestles had been rapidly erected to replace Bridge #6 and the massive Conemaugh viaduct. The stone viaduct had collapsed at the height of the flood. Its temporary trestle was built by June 12th (*Eng. News* 1889a). A new masonry viaduct was completed before year's end.

No doubt the swift mobilization by Pitcairn of all his forces saved lives by quickly bringing aid to the Conemaugh Valley. He deserves credit for this at least. Notably, the PRR for a time transported relief supplies to Johnstown at no cost. But Pitcairn had another motive as well – the income from trans-state commerce on his line was at a complete standstill until the railway was restored.

5.2   Eyewitness Accounts from the South Fork Dam

Eyewitnesses are essential sources of information to reconstruct events. But as is well-known to crime detectives, their accounts can be inherently unreliable, depending on the people, their frame of mind, time of day, and the nature of the events. The accounts of John G. Parke Jr. include newspaper articles and a detailed letter he wrote to the ASCE committee less than 3 months after the flood. Although he was inexperienced, and had not finished his engineering studies at the University of Pennsylvania, Parke was the only engineer who witnessed the dam breach and therefore the committee placed great reliance on his observations. Their report transcribed his letter in full as the best first-hand account they had obtained. Some other people also saw the dam failure, and parts of their testimony to the PRR are included here. Notes about John Parke's life and career after the Johnstown flood are included in the chapter of biographies of the ASCE engineers.

5.2.1   John Parke's Letter to ASCE's Investigation Committee

The letter from John Parke gives us a detailed record of events at the dam on the day of the flood, although there is also testimony from observers given to the Pennsylvania Railroad. Parke himself was not interviewed by the PRR.

As resident engineer at the Club that spring, some survivors in Johnstown may have held him or other Club workers, such as "Colonel" Unger, responsible for the tragedy. I found no evidence that Parke himself received threats. But he posted his letter from afar - Americus, Georgia on August 22nd, 1889. His work at the Club was clearly finished and it may be that influential family members arranged for him to at least temporarily remove himself from Pennsylvania. The letter Parke sent to the ASCE Committee is detailed and professional, illustrating keen powers of observation and his overall sense of peril to the dam as the lake waters rose. But it is difficult to know how much that "sense" of peril resulted from his reflection on the disaster and discussions with others during the nearly 3 months that passed until he penned his letter to the Committee.

Parke's letter is given here in total, quoted from ASCE's 1891 report (p. 448–451) and interspersed with my commentary in italics. Additional paragraph breaks were added for clarity; one of Parke's paragraphs was 3 pages long! Parke addressed his letter to Max Becker, the railroad engineer who was the standing ASCE president in 1889 and also one of the investigators.

M. J. Becker, Esq.,

President Am. Soc. of C. E.:

DEAR SIR, – Being requested by your Committee to give an account of the destruction of the South Fork Dam, I will do so by narrating my experience of the affair and the few observations I made at the time.

On the evening preceding the destruction of the dam, to the best of my recollection, we had many evidences of an approaching storm, and when it grew dark, we had a violent wind storm, and the tree tops about the house were moaning and creaking unusually, but no rain fall. About 9 o'clock I had occasion to go out of the house a short distance and noticed that the board walk was wet and that there had been a slight rain-fall, but it was not raining at the time, and the sky was much brighter and evidently clearing off, but there was still a high wind blowing. These sudden, violent wind storms, very often accompanied by heavy but brief rain-falls, were customary in that mountainous country, as I had noticed during my 2 months' location at the lake previous to this time.

I retired shortly after 9 o'clock and slept very soundly, awaking once towards morning and hearing a heavy rain. [*4 days after the flood Parke had told a Pittsburgh reporter that he did not hear the downpour - see paragraph following this letter*] When I awoke at about 6.30 on the morning of the 31st, I found it very foggy outside, and on going out, found the lake had risen during the night probably 2 feet, and I heard a terrible roaring as of a cataract at the head of the lake, about a mile above the club house where I was staying [*for a "roar" to be heard at a distance of 1.6 km suggests the inflows to the lake at that time may already have been approaching a peak discharge*]. After eating breakfast and returning to the shore [*the shoreline was very close to the clubhouse*] I found the lake had risen appreciably during our absence, probably 4 or 5 inches [*in a half hour or so; this estimate suggests a rate of rise around 8–10"/hr sometime between 6:30 A.M. and 7:30 A.M.*], and with difficulty secured a boat, and with a young man who was employed on some plumbing work at the cottage, I rowed to the head of the lake to see the two streams that were pouring into the lake with such an unusual roar [*these two streams were the main stem of the South Fork of the Little Conemaugh and what is now called Laurel Run, which has a tributary named Mud Run*]. I found that the upper one-quarter of the lake was thickly covered with debris, logs, slabs from sawmill, plank, etc., but this matter was scarcely moving on the lake, and what movement there was, carried it into an arm or eddy in the lake, caused by the force of the two streams flowing in and forming a stream for a long distance out into the lake. The lake seemed very high when I reached the head, for we were able to row over the top of a four-wire barbed fence which stood near the normal shore line and we rowed for 300 feet across a meadow which was covered with water, for it was very flat and a rise in the lake of a foot, covered a large area. I did not get very near to one stream (the Muddy Run) [*he would have had to cross the tremendous inflow at the mouth of the South Fork to reach the mouth of that stream /AKA Laurel Run*] but could see the volume it was pouring in by its current in the still water, but we did go up the shores of the South Fork Creek and found it widely over-running its banks. In its normal condition it is about 75 feet wide and barely 2 feet deep – many places not that deep, but varying as every mountain stream does. But on this day it was a perfect torrent, sweeping through the woods in the most direct course, scarcely following its natural bed, and stripping branches and leaves from the trees 5 and 6 feet from the ground. We tramped through the fields adjoining the woods in which the stream was boiling, for a half mile above its mouth and I could appreciate its volume and force, for I was familiar with the region and saw where the stream covered a portion of the township road for a depth of 3 feet, which is never covered except in floods.

Returning to our boat, we found it almost adrift, the water having risen during our absence. We rowed to the club house and found the water had risen at a wonderful rate during our row to the top of the lake. I had been thinking of the dam and was not surprised when landing to be told that the water was nearly over the dam and that men and a plow were needed there. So taking a horse from the stable, I rode to the breast and found Colonel Unger, President of the South Fork Fishing and Hunting Club at work on the dam with a number of Italian laborers (that we had employed on some sewerage work); there were about sixteen of them. Half of them were cutting a ditch through the shale rock at one end of the breast [*this was on the western abutment, where a parking lot exists today*]. The shale was so tough that they could not cut it more than about 14 inches deep and about 2 feet wide, but when it was cut through to the lake, the water rushed in and soon made it a swift stream, 25 feet wide and about 20 inches deep, but the rock was so hard that it could not cut it any larger than this. [*The "hardness" was due to the abutment having been excavated by the State workers to make an auxiliary spillway – the soft soil and regolith above the bed-rock had been removed*].

Previous to this [*ditch*] being opened and shortly afterward, I made two observations of the height of the water, and the lake in the hour had risen 9 inches [*a useful observation of lake rise at that time; he was quoted elsewhere as saying the lake had risen at 8 to 10 inches per hour*]. During the digging of this ditch, I rode back and forth over the dam directing the laborers. We had a plow at work throwing up a furrow, and thus raising a temporary barrier or breast to retard the water flowing over the dam, should it reach that height which it was gradually doing. I noticed that the waste-weir proper was discharging to its full capacity, and that there was no drift or other matter to clog it, except a road bridge supported on small posts which were apparently offering but little resistance as the weir [*main spillway channel*] was narrower by about 15 feet at 100 feet from its mouth, and this contraction compensated for the resistance to flow offered by the bridge supports. There was probably 7 feet of water in the weir at the time. There were some iron screens between the foot of each post on the outer row of the bridge supports, but they were but 18 inches high and could not have been removed, had we wished to, owing to the depth and velocity of the water.

The water in the lake rose until it was passing over the breast, notwithstanding that the lake had then the two outlets (the waste-weir and the one cut by the laborers). The breast was slightly lowered in the center and the water washed away our temporary embankment thrown up by the plow and shovels, and the water was passing over in many places in a distance of 300 feet about the center of the breast; the men stuck to their task and worked until the water was passing over in nearly one sheet, and then they became frightened and got off the breast. I saw what would be the consequence when the water passed over the breast and rode to South Fork Village and warned the people in the low lands there, and had word telegraphed to Johnstown that the dam was in danger. [*by this Parke implies it was his idea to ride to South Fork to give warning, but his boss Col. Unger would have sent him because Unger was closely supervising all work at the dam; nonetheless, there was risk to Parke had the embankment failed while he was on the road between the dam and South Fork, and later when he traveled across the dam while it was being overtopped*] The people in South Fork heeded the warning and moved out of their houses.

When I left South Fork to return it was just twelve o'clock noon, and the water had been flowing over the dam for at least a half hour [*therefore overtopping of the dam crest began at 11:30 am or earlier*]. I rode back up to the lake 2½ miles through the valley and found the men had torn up a portion of the flooring of the waste-weir bridge and were endeavoring to remove the V-shaped floating drift guard that projected into the lake. It was a light affair and was built to float on the surface of the lake and catch twigs, leaves, etc., and prevent their clogging up the iron screens spoken of above. I crossed the breast at this time [*he later said he crossed it on foot, around 1 pm; his ability to cross the dam crest at this time, after overtopping for more than an hour, suggests that the sag in the dam center was probably no more than ½ to 1 ft*] and found the water was cutting the outer face of the dam, but not as badly as I feared it would, its greatest effect was on some portions of the roadway which crossed the breast where the roadway had been widened on the lower side by the addition of a shale earth or disintegrated shale, upon which the action of the water was instantaneous, but the heavy rip-rapping on the outer face of the dam protected this wash and the water cut little gullies between each of the large stones for rip-rap.

I did not stay on the dam when it was in that condition, but went on to the end of the dam and found that over its entire top it was serried by little streams where the water had broken through our little embankment [*made with plough and shovels*] and was running over the dam. I went on to the new waste-weir we had cut and found it carrying off a great volume of water and at a great velocity. I with difficulty waded it and found it was up to my knees or 20 inches deep. I felt confident that nothing more could be done to save the dam unless we were to cut a wasteway through the dam proper at one end and allow it to cut away in but one direction, and that towards the center of the dam [*Parke probably got this idea from reading a June 15th article (p. 551–553) in Eng. News (1889b) which said "A man of great resolution, self-confidence, and self-sacrifice might have diminished the disaster when he saw it was inevitable by cutting the dam at one end....This would have greatly decreased the rate of erosion, since there would have been only one side to erode, and that could have been in the more solid old work."*], but this I would not dare to do, for it meant the positive destruction of the dam, and the water at the time was almost at a stand, owing, without doubt, to the large increase of outlet by the overflow on the breast, and I hoped that it would not rise, but yet expected it to rise for it had been raining most all of the morning, and consequently we had more water to expect.

I hurried to the club house to get my dinner and to note the height of the water in the lake, and found that it was a little over a stake, that from my level notes of a sewer I was constructing I knew was 7.4 feet above the normal lake level [*consistent with a dam crest lowered 3 ft by the Club, and an overtopping depth at the low dam center of 0.4 ft*]. I returned to the dam and found the water on the breast had washed away several large stones on the outer face, and had cut a hole about 10 feet wide on the outer face and about 4 feet deep, the water running into this hole cut away the breast in the form of a step both horizontally and vertically, and this action went on widening and deepening this hole until it was worn so near to the body of the water in the lake that the pressure of the water broke through, and then the water rushed through this trough, and cut its way rapidly into the dam at each side and the bottom; and this continued until the lake was drained. I do not know the actual time it consumed in passing through the breach, but it was fully 45 min. [*most modern compilations of dam breaches cite his 45 min as the actual time to drain the lake; Parke clearly stated this as a <u>minimum</u> estimate*] It did not take long from the time that the water broke into this trough until there was a perfect torrent of water rushing through the breast, carrying everything before it, trees growing on the outer face of the dam were carried away like straws. The water rushed out so rapidly that there was a depression of at least 10 feet in the surface of the water flowing out, on a line with the inner face of the breast and sloping back to the level of the lake about 150 feet from breast, exactly similar to water flowing through a rectangular sluice-way in the side of a trough with the water level far above the bottom of the sluice-way. When the lake was drained there still remained in the bed of it a violent mountain stream 4 or 5 feet deep, with a swift current the combination of the two streams already alluded to from the head of the lake and the many little streams from the adjacent hills, which streams were all overflowing their banks, this stream in the bed of the lake showed no signs of diminishing in volume until late in the following day, and was impassable with a boat for several days.

I need say nothing of the character of the dam, for it is open for inspection of those far more able to express an opinion than I. But there is one thing I want to impress on every one's mind, and that is, that the dam did not break, but was washed by the water passing over it from 11:30 o'clock A.M. until nearly 3 P.M. until the dam was made so thin at one point, that it could not withstand the pressure of the water behind it, and the water once rushing through this trough nothing could withstand it.

Hoping this report will be of some service to your Society, and placing myself at your disposal to answer any inquiries as to the destruction of the dam or any points that I have not developed, believe me,

Respectfully yours,

JOHN G. PARKE, Jr.

AMERICUS, Ga., August 22d, 1889

The night before the dam breach, Parke Jr. had been sleeping at the clubhouse and wrote in his August letter to the committee that he "…slept very soundly, awaking once towards morning and hearing a heavy rain…" His letter was composed 12 weeks later and posted from Georgia. The letter varied in one fact from an account he gave to a reporter from a Pittsburgh newspaper just 4 days after the flood. Parke told the reporter that "It rained very hard Thursday night I am told, for I slept too soundly myself to hear it. ..." (*Gazette* 1889). This statement a few days after the disaster is more believable, that Parke Jr. himself did not hear the pre-dawn downpour. I do not suggest that he gave an intentional misstatement to the Committee. The different versions may simply reflect human nature; as an eyewitness to an extraordinary incident he would have been asked many times to recount the events he saw that day at the lake. After enough retelling he may have convinced himself that he actually awoke before dawn and heard that downpour, and thus relayed an account that was instead overheard from others at the clubhouse that day.

Parke told the same reporter that the dam breached at 3:00 P.M., and in his letter told the Committee that it sustained overtopping "…until nearly 3 P.M." This entirely fits with the estimate by Kaktins et al. (2013) based on stoppage of the railroad station clock at South Fork and the time needed for the flood wave to travel from the dam to the station, where it knocked the building and its clock mechanism off plumb. The dam failed between 2:50 and 2:55 p.m.

Parke told the *Gazette* (1889) that "It was a terrible sight to see that avalanche of water go down that valley already choked with floods. Col. Unger was completely prostrated by it and was laid up at the club-house sick from his experiences." Parke had also given a reporter the names of some others who were present at the clubhouse on the day of the flood. The Clubhouse registration book gives no clue about that because of the pages missing from the bound ledger. Besides Unger and various laborers, also present (according to Parke) were DeWitt C. Bidwell and family members; J. J. Lawrence; George Shea and his brother; Louis Irwin and son; and James Clark. They reportedly went on to Cresson, PA on the Sunday after the flood.

5.2.2 Other Eyewitnesses of the Dam Breach

Club grounds superintendent W. Y. Boyer also witnessed the 1889 dam breach (*Tribune* 1889 p 1).

> …On Thursday (the day before the breaking of the dam) it commenced raining about 4 P.M., but did not rain very hard until about dark, when it commenced to blow and rain, and all that night it rained very hard. It not only rained, but it poured, and in the morning when I got around I told the guests that I thought the dam would run over that day, of course not being sure at that time. The water rose ten inches an hour.
>
> I left for South Fork shortly after breakfast [driving the Bidwell's], and returned about 10 o'clock, when I found that the President of the Club [Unger], who was up here at the time, had ordered all the men to help move the dam. Everybody—twenty-two Italians and many neighbors—worked hard trying to save the dam, but in vain; the water finally commenced to go over the dam and began to wash from the outside.
>
> It would wash a kind of puddle and then some of the earth would give way, and keep on in that way till it had a channel through, when you may know it went with a crash, where nothing could escape that was in the way. Trees four feet in diameter, roots, branches, everything went before it like toys, and before it reached South Fork [village] it was a dam of trees, and not of clay and stones as before; and, you may believe, I don't want to see another breaking of a dam, and I never dreamed that I should be an eyewitness to such a thing when I left the quiet little village of Garfield, Ohio…

McLaurin (1890) gives additional accounts from eyewitnesses John Rorabaugh, whose farm adjoined the lake, and George Gramling. Rorabough had lived near the South Fork dam for 45 years. He said that the dam crest was lower on the southwest side. That is also where the trench was dug on the day of the flood to "let off the water." We now know that this "lower" end of the dam was in fact an emergency spillway. Rorabaugh thought the dam before the flood was about one foot lower in the center. He stated:

> In the morning it was raining hard. Thinking the water in the reservoir would rise to a great height, I went down to the breast. The water was then rising ten inches an hour. A gang .... was put to work at the south side of the dam to make an opening, and did succeed in letting some water out. The embankment was hard to cut, and little headway was made. The water continued to rise. At one o'clock, when I visited the dam a second time, the water was running over the breast. I soon went home, returning in an hour. About three o'clock a break occurred in the breast of the dam, and the whole mass of water rushed with a tremendous roar down the valley. At the top the break was about three hundred feet wide and it sloped down to about two hundred, below which another break occurred about twenty-five feet wide, through which the stream now runs. I have been a resident of the reservoir neighborhood since 1844 and know about the construction of the dam. When the State first built it the breast was made entirely of clay, packed in layers, backed with rip-raps of stone. The Railroad Company made no change in the dam. When the Pittsburgh people got hold of it they began to make some additions to the breast. They hauled stone and patched up a break, and raised the breast [*they were filling the large breach*] and widened it with stone and earth. When Colonel Unger saw the condition of the dam — some time before it broke — he remarked that if it withstood this flood the association owning it would put it beyond all possibility of danger in the future. But it didn't hold, and when the Colonel saw it go he, realizing the awful consequences of the break, became so ill that he had to be assisted to the hotel [clubhouse]. (McLaurin 1890 p 55)

George Gramling owned a sawmill and gristmill on Sandy Run, southwest of the South Fork dam. That stream entered the South Fork below the dam. His mills were powered by water from a small dam that had failed at seven in the morning (McLaurin 1890 p 55). This led:

> ...Mr. Gramling to think the big dam would go also. He and E. S. Gramling, Jacob G. Baumgardner and Samuel Helman started about 8 o'clock for the lake. When they arrived the water was six feet from the top of the breast and rising about a foot per hour. Toward noon Mr. Gramling went home for dinner and returned in two hours. Crossing on the bridge below the dam, he went up to the top and walked on the bridge over the waste-weir. The water was then running over the lowest portions of the crown half way up his boot-leg. He remained until the breast broke and the water started down the valley. The water, as it tumbled into the stream below, gradually washed the embankment away until it was not more than half its original thickness. A short section in the middle of the dam gave way, increasing as the waters swept through until the gap was a hundred yards wide. Had this gap been made all at once at the first break the flood must have been even more disastrous. It was 15 min from the time the dam broke until the great bulk of the water was discharged, if Mr. Gramling's estimate be correct, and it accords closely with others.

Gramling's time for the "great bulk" of the lake to drain is greatly underestimated, and shows how time can fly when witnessing a traumatic event. It is also true that the lake level would have dropped ~4 meters during the first 15 min following the general dam failure. That rapid lowering would have exposed a great expanse of lake bed.

5.3   John Parke's Ride to South Fork

In at least one account John Parke suggests that he rode to South Fork on his own volition. The June 4th edition of the Gazette (1889) quotes him as saying "By 11:39 I had made up my mind that it was impossible to save the dam and getting on my horse I galloped down the road to South-Fork to warn the people of their danger." But Col. Unger was on the dam directing all the work to try to save the dam. It is unlikely Parke simply rode off without being sent by Unger, or at the least advising Unger and getting permission. Parke worked directly for Unger and in his letter to ASCE makes reference to Unger's instructions for emergency work at the dam. But Unger would clearly have sent Parke on that mission and deserves the credit for doing so. In the absence of a working telephone at the Clubhouse, Park rode on horseback to send Johnstown and the other towns a warning.

Photographs of the South Fork dam taken before the flood show wooden poles erected along the crest. These did not power electric lights; they carried wires that by the late 1880's were used for a telephone service.[6] It is unfortunate that the telephone at the Clubhouse was only used during the busy summer season. Unger himself could have more fully described conditions at the dam for the operator in South Fork, resulting in detailed messages for officials in Johnstown. Perhaps such a personal warning from Unger would have been spread to more people. But Parke will always be recalled as the young engineer who rode down the valley to try to warn South Fork, Johnstown, and the other towns. One of Parke's obituaries, from Monessen where he had lived and worked for many years, contains blatant false hyperbole about his ride to South Fork:

Monessen Engineer Gave Warning to Johnstown

SAVED THOUSANDS

Raced Horse Through City's Streets With News of Peril.

Special to the *Pittsburgh Post-Gazette* – MONESSEN, Pa., Jan. 30. --- The "Paul Revere" of the Johnstown flood is dead. The man who is credited with spreading the alarm through the streets of Johnstown on horseback before the flood struck the city in 1889, died at his home here at the age of 67....

John Parke never rode to Johnstown that day, or even west of South Fork. He had been sent by Col. Unger from the dam to South Fork to warn the people there and telegraph a warning to points west, including Pitcairn in Pittsburgh. The warning was read at each relay point, even carried by foot where the lines were down. As Parke rode to South Fork we can wonder if he knew that two of his classmates from the University of Pennsylvania were in Johnstown, including one who had been a class President. One would live and one would die. After arranging for the warning to be sent Parke quickly rode back to the Club where he soon witnessed the catastrophic dam breach.

---

[6]The first commercial telephone exchange in the U.S. opened in New Haven, Connecticut in 1878. Telephone service between New York and Boston opened in 1884.

There was a fictitious claim of a heroic horse rider who tore through the streets of Johnstown to warn the people. McCullough (1968) reviewed this claim and its various sources. In one, the fable appears in a poem printed by Connelly and Jenks (1889), "The Ride of Daniel Peyton, the Hero of the Conemaugh." Their vaunted hero perished, "…a mighty wave, blacker and angrier than the rest, overtook horse and rider and drew both back into the outstretched arms of death." Survivors of the flood tried to track down this "hero" but found the story without merit. Like McCullough I found no evidence of a Dan Peyton or his ride. If he existed and was known to so many why did no one report him missing - his name is not among the victims. There were four Peytons of Clinton Street who died in the flood, Julia F. (13), Marcellus K. (16), George A. (19), and 65-year-old John W. Peyton. None were named Daniel, and all four were interred in Grand View Cemetery.

But it is certainly true that early writers of the flood, like Connelly and Jenks, and another opportunist named Johnson, stretched for hyperbole – it enhanced the clamor for their books.

5.4 Witnesses along the Flood Path

Photographers came to the flooded areas as soon as they could get transport, including by horseback. Their black and white images give us a surreal understanding of the destructive power of floods. It is all the more fascinating to read about the experiences of people who saw the flood wave barrel down the valley of the Little Conemaugh. In many cases the witnesses themselves had to run for their lives.

Their stories come down to us through the PRR testimonies. Robert Pitcairn had directed that this record be made, in particular reference to the loss of life and property on the passenger trains. His intent was to gather information about whether the Company was legally liable for those losses and to help prepare a defense to any actions that might arise. The testimonies were prepared in a question and answer format by attorney John Henry Hampton at his Pittsburgh office, beginning on July 15, 1889, a month and a half after the flood. Most of the witnesses were railroad employees, but statements of some others were taken. Pitcairn himself gave extended testimony, as did his Assistant Superintendent, Michael Trump.

John Hampton was the chief legal advisor for the Western Division of the Pennsylvania Railroad Company (Fig. 5.1). He held this position for 34 years from 1857 until he died in April, 1891. He was said to be a delightful companion, always interesting, kindly, and genial. His friend and law partner John Dalzell said of him, "He stimulated thought in others; pointed the way, where, before he pointed, no way appeared. He had a lofty conception of his profession; an instinctive contempt for those who would debase it. He despised a mean action and a mean man. As a lawyer he lived up to his conception." "He was a great fisherman, and loved the woods and fields and all the sights and sounds of nature." (PRR 1892).

Since the testimonies were taken more than 6 weeks after the flood there is the possibility that some witnesses consciously or unconsciously added the observations

Fig. 5.1 Images of John Hampton, Pennsylvania Railroad attorney. (Credit: PRR 1892)

of others but reported them as their own. In reviewing the transcripts it is important to consider how the responses were shaped by the specific questions posed by Hampton. Some heard open-ended queries that encouraged broad answers, while other witnesses got pointed questions to elicit narrower responses. Pitcairn himself did not miss the chance to make some questionable statements about the earlier history of the South Fork dam and the Club. A few of those statements will be examined further.

Testimony of DeWitt Bidwell - at South Fork Station DeWitt Bidwell (1828–1900) was a Club member who owned a company in Pittsburgh that sold explosives to the mining industry. He was also involved in banking and served as a representative in Pennsylvania for the Dupont Corporation. On the morning of the flood he and family members rode from the Club down to South Fork in a carriage driven by Boyer. They intended to catch a train but found all were delayed. Bidwell saw John Parke ride into South Fork, and several hours later witnessed the flood wave as it struck that village.

> Well, it came in a large volume with a good deal of debris. All this time it was raining, and I was standing at the [*South Fork*] station house, watching the creek, and going outside with my umbrella and standing in the rain to see what was going on outside, and about half past 2 o'clock, I had a conversation with Mr. Dougherty [stationmaster] about the passenger train standing in the bend of the road at the tower. The New York and Chicago Limited came there with a train of six cars in the morning, and Mr. Dougherty got the engineer to run his train up past the station; and as that train went up above the station, a freight train standing in at the left on the switch near the Supply Company store, started up also and got up out of the way. There was a freight train standing on the right hand siding down at the coal works that started, and had very little steam on the engine, and the locomotive just got over the iron bridge when the flood struck it, and carried off the train. There was nothing left there but the locomotive.

C. P. Dougherty – PRR Station Agent, South Fork Dougherty had worked for the PRR since 1864, serving at South Fork since November 21, 1887. Pitcairn had high regard for him, calling him one of "our oldest and best freight conductors" and "a first class railroad man." But when the Company started using color vision tests, Dougherty proved to be color blind. Given the safety importance of colors in train signals he could not be promoted to yard master. Pitcairn gave him the Agent job at South Fork until a better job came available.

In Dougherty's testimony, from 1:00 pm onward he thought water levels in both the North and South forks of the river were "…lowering a little which renewed a hope that it was also lowering at the dam…" He was standing 600 or 700 feet east of the station. From there a line of buildings blocked his view up the valley of the South Fork. His first warning was from a man yelling behind the station who was trying to direct attention up the stream. Then he saw people running around the corner and knew there was trouble in that direction. Dougherty started to run for the station and his company home, which was just west of the station. He thought the dam broke "…about 2.45 or 2.50." As he was running he got to the station and could see the flood coming down.

> It had the appearance to me of being about 40 feet higher than the level of the roadbed; I mean the current coming down the South Fork stream … It first gorged the valley of the Conemaugh west of the point where the two streams unite, and then struck directly across to the opposite side, and seemed to form a swirl from the opposite side of the mountain, and ran up the North Fork [*east – reversing its flow*] at a rapid gait, and continued on up a considerable distance before it backed the water up at the station …. It seemed to be much higher in the North Fork for some time before the back action came towards the station. That gave those people living along there time to escape, and in my opinion was the means of them all getting out of the way. While it was in this position, they had the chance of the delay in the back water coming, and they of course escaped to high ground. It [*the reverse flow up the North Fork*] was probably half a mile east ---
> The way I understood it, the amount of stuff the water was pushing when it came down the South Fork, large timber, and big hemlock trees end over end, and the force it struck the mountain with, gave it that surge up the North Fork, and it apparently passed this locality where those buildings were, and the station, before it overflowed that district… In my judgment, it was traveling at least 15 miles an hour.

Dougherty described how his station house lifted off its foundation and floated a small distance *upstream* before drifting back and settling near its original spot. Surprisingly, Bidwell never mentioned that the station where he had been standing had floated like a boat! Degen and Degen (1984 p 22) show a post-flood image of the shifted South Fork station, standing by the tracks with a severely warped platform, and nearby utility buildings tumbled over as if they were toys.

Emma Ehrenfeld - Telegraph Operator, South Fork Tower Ehrenfeld worked as an operator for the PRR. She had gone on duty at 7 a.m. on the day of the flood. She had received the warning message about the dam, from a man sent by John Parke. It is unclear why Parke himself did not deliver the message. In mid-afternoon she witnessed the flood wave:

> …the engineer and the conductor of the 1165 [*train*] were in the office at the time, and it seems were looking out of the window; I was sitting with my back to the window; and they said, 'Look at the people running! I wonder what's wrong.' I looked up and went to the window, and just then, it seemed the cry arose that it was coming, and he looked out of the window and said something about the reservoir going, and he and the conductor started

down stairs. I then went to the window and looked out and saw people running, and some were screaming, and some hollowed for me to come, and I looked out of the window on the side of the river, and saw it coming … It just seemed like a mountain coming, and it seemed close; of course, I don't know just how close it was, but I knew I must go if I wanted to get out, and I started and ran down the stairs without waiting to get my hat or anything; and there is a coal tipple about opposite the office, and I ran down across the track, and up those steps. It was a very short time, not more than two minutes until the office [*tower*] was taken.

P. N. Pickerell - Telegraph Operator, Mineral Point Tower   Pickerell was sitting in the tower and suddenly heard a roar. He looked up the track and saw trees and water coming, then climbed out a window onto a tin roof. Driftwood came around the curve, the channel engorged with water running over the bank. He heard voices but couldn't see anyone. More drift come down the river and then he saw a man named Christ Montgomery standing on a house roof. Christ saw and recognized Pickerell and yelled that Mineral Point was all swept away and the people swept away, and his own family gone. Pickerell yelled encouragement to him. Further downstream the floating roof struck the shore and the man crawled up the bank and escaped. Pickerell then turned and saw "…a regular mountain of water coming." He ran out of the tower and over the track onto a high bank. He had taken the time to grab his hat and coat as it was raining hard. He saw the flood wave nearly catch Montgomery, who escaped the water a second time. Pickerell feared for the lives of his own family in Mineral Point. They had survived but he soon learned of relations lost down in Johnstown. Pickerell later estimated the flood water reached a height of 24 ft. at his office tower.

Michael Trump – At AO Tower   The flood wave surged downriver from South Fork until it reached the Conemaugh viaduct, where water temporarily backed up. Collapse of the viaduct sent the renewed flood wave down the valley toward Mineral Point, just 1.2 km away. Beyond Mineral Point was Bridge #6, the AO tower, and East Conemaugh. Michael Trump had pushed his train past Buttermilk Falls and around the big curve there. He and his crew had seen debris in the river that they thought must be from the dam. Trump told the conductor to stop the train before reaching the AO tower and they all got out and began running toward the tower. In the PRR testimony Trump described what he saw next:

> Before we reached [*AO*] there however, the telegraph poles along the river commenced to go, the wires had pulled the tower out of plumb, and the operator got out. We ran up to him, and asked him whether he could get in and send a report. He said he had already reported the matter, and Mr. Wierman and Mr. Webb then ran on up towards bridge 6 and I remained in the vicinity of the tower. The water seemed to rise about a foot at a time for a while. We would pick out a land mark, and look at something else, and when we would look back for the land mark, it was gone; it was just rising that rapidly; jumping up a foot at a time. The thing got so terrific that we began to hunt a hill for ourselves. We got on the track near a hill where we could get up easily. The water commenced to hit the trees and snap them off, the telegraph poles were twisting the wires in every direction. We then moved on up toward bridge 6, and we noticed the water running through the deep cut from 15 to 20 feet high before the bridge went out. Then I heard Mr. Wierman and Mr. Webb call out that the bridge had gone. They got there in time to see it go. After that, there came a wave along that seemed to be six or eight feet high, and then the water kept about the same level; finally another wave came down; this final wave seemed to be about ten to twelve feet high above

the other water. Then, after this wave, the water seemed stationary for a while, and finally receded. We then decided to go back towards [East] Conemaugh, and we got on our train, and when we got around Buttermilk Falls, the track was all gone. The water swept around there next to the mountain, so that we were entirely cut off. By that time, it was about five o'clock.

Trump had reported several successive waves or crests on the river, one of which he called the "…final wave [that] seemed to be about ten to twelve feet high above the other water." The initial flood wave that Trump saw would have been caused by water pouring across the narrow neck of land at the Conemaugh viaduct and from floodwater passing under the bridge before the arch became clogged with debris. The "final wave" he saw must have been spawned by the collapse of this structure.

Trump sent two men up into the hills to look for horses and get a message to Ebensburg. Two other men, Mr. Reinhart and a crewman, were sent over the ridge to East Conemaugh to report what they had seen. Those two men made it there and got back to Trump after dark. Trump and his people stayed in place until morning, then went over the hill to East Conemaugh. They eventually made their way along the north bank of the river to Johnstown, or what was left of it. Trump had seen for himself the destruction at East Conemaugh, including a number of corpses. In Johnstown he encountered Agent Deckert again, and found he could not report to Pittsburgh because the telegraph lines were down to the west. Trump's group then made their way along the river to Sang Hollow and beyond to "SQ" tower. They went a few more miles and got across the river by boat. From New Florence Trump was finally able to send a detailed message to Robert Pitcairn.

William Adams – PRR Engineer, on Hillside Across the River from East Conemaugh Station Adams had been on one of the Day Express engines and said he had not been told of any warnings about the condition of the South Fork dam. He had gone to his house which was on low ground to help his wife move up the hillside for safety. Before crossing to his house on the south side of the river he noticed the flood level had dropped a little. He moved carpet and some furniture to the second floor. Adams then went out to check on the river and heard an engine whistling upriver from the roundhouse. That whistle had been tied down by John Hess,[7] the engineer on the work train east of the station, who was driving back down the valley. A continuous engine whistle served as a warning to railroad men. Fast as he could, with the help of another engineer, Adams carried his disabled wife on a chair up the hillside to higher ground. In his testimony Adams said "We just got over the old Portage Railroad when the houses were all going." The flood wave "…just looked like a great body of high water, as high as a house, 30 feet, full of timber, logs, trees, and everything."

Adams had lived at the same place years earlier when the dam partly breached in 1862. He recalled that earlier breach had "…made the water pretty high; just a regular high water." The channel was "…bank full, but it didn't do any damage at all."

---

[7] Hess survived the flood and gave a brief account in the PRR testimony to John Hampton. There is no question that Hess's continuous signal saved lives that day.

S. E. Bell – Conductor on the First Day Express, East Conemaugh Bell reported there were seven cars in his train. He had been informed by Mr. Walkinshaw that the dam upstream was liable to go, and that was given as the reason for moving the train to a higher siding. Apparently Walkinshaw planned to notify the crews if the dam did break. By that time there were rumors among the passengers about the dam. Bell had briefly stepped into the rear car of the mail train to get out of the rain. Just then a work train upriver began whistling like an alarm. Bell jumped out of the car and saw the flood wave coming. "It looked to me like a wall about thirty feet high; I don't suppose it was that high, but it looked like that to me, and I thought the only salvation was to get to the hill." He ran to his Day Express and shouted warnings to the passengers who remained and then headed for the hillside. He later estimated that more than 40–50 houses there were swept away.

Alfred H. Brown – Train Passenger One of the Day Express passengers who survived wrote of his experiences in East Conemaugh that day and of his subsequent journey with other survivors (Brown 1889). Alfred Brown was the private secretary to the chairman of the Central Traffic Association of Chicago. He had left Chicago on the Day Express #6, arrived at Johnstown at 10 a.m. the next morning, then pushed on to East Conemaugh by 10:30 where the passengers were told the train would have to wait for a bridge repair. To stretch his legs Brown got out of the train and walked up and down along the tracks and noticed the river was so high it was eating away at the embankment 150 yds. west of the station house. He reported it had been raining the whole time but by about 3:00 p.m. the rain was "…commencing to get a little heavy."

Brown went back to his railroad car, the "Aragon," and read a book for about an hour. He then heard shouting, looked out the window and saw people running for the hillside. Brown calmly put on his hat and coat, grabbed his umbrella, and on stepping down from the train to the platform he was told the South Fork dam had given way and the flood was almost upon them. He ran about half way to the hill and stopped to look upstream, later describing the first appearance of the flood – it was like a mountain stream running around the station. Then a two-story frame house suddenly rose into the air about 10 feet and crumbled into pieces against several smaller houses. Then abruptly Brown saw "… a perfect avalanche of houses, water, trees, freight cars…" rushing toward where he was standing, more than 100 yards away. He froze for a moment then again raced for the hill. Turning, Brown saw the flood rush by the very spot where he had stood moments before:

> The roar of the elements was simply deafening and seemed more like the roll of thunder than anything I can think of.

Brown (1889) wrote that the flood lasted about an hour. He credited the placement of the trains and the formation of a small debris dam for making it possible for many passengers to escape with their lives. Brown and another man got overnight lodging with a track repairman named Warfel and his wife. During the night three of the Pullman cars that remained caught fire and burned, including Brown's "Aragon" car. The next morning Brown walked west from East Conemaugh to beyond Woodvale. He wrote that only 7 or 8 houses remained there, high up on the hillside. All others had been swept away with great loss of life. He saw a constable guarding a safe that had been washed a mile downriver. The safe reportedly belonged to a druggist in East Conemaugh.

Brown left East Conemaugh on June 1st at 1:45 p.m. on a farm wagon, bound for Ebensburg. The ride was rough, boards having been nailed across the sides for seats, and the road itself was rocky and bumpy. With 10 other passengers on the wagon they got to Ebensburg by 9 p.m., then boarded a train for Altoona and arrived nearly an hour after midnight. Brown had not heard the message that the PRR had arranged accommodations for stranded passengers. He spent the night in a private home generously offered.

During the jolting ride to Ebensburg, Brown struck up a conversation with a passenger named Mrs. M. J. Blaisdell from Minnesota who described her own narrow escape from the flood wave. She was one of the last to get off the express train and found her way blocked by a disabled man. He heard her anguish and threw himself aside so she could escape. She ran for the hill, falling several times, and was caught in water up to her waist. She was pulled out by a railroad brakeman.

Alfred Brown eventually arrived in Jersey City at 7:35 a.m. on June 7th, having taken a week and 19 h to travel from Chicago to New York. One further note about his letter is that it contains some wildly exaggerated claims about the death toll from the flood, and also evidence of his cultural and racial biases, particularly against immigrants, that was sadly all too common in that and other eras.

Brown's story about Mrs. Blaisdell was corroborated by the PRR testimony of D. T. Brady, the front brakeman on the mail train at East Conemaugh. He was the one who helped her to high ground and she later gave him her full address: Pelican Rapids, Otter Trail County, Minnesota. In the testimony her last name was written "Blazdell." Brady reported that his only knowledge about hazard from the South Fork dam that day was from rumors among the railroad men about a message from the telegraph tower.

5.5 Telegram from the President

In the aftermath there was immediate concern about an unnamed railroad passenger, voiced from the highest level of U.S. government. President Benjamin Harrison, writing from the Executive Mansion, telegraphed:

General D. H. Hastings,

Please send me at once full information concerning two trains supposed to have been near Johnstown. Give me names of passengers. Answer promptly please.

Benj. Harrison.

President Harrison may have been short in stature but his words were clear and strong through the telegraph wires. Twice married, Harrison was the 23rd U.S. President, the grandson of William Henry Harrison, the 9th President. After doing some research about him and his family it remains a mystery who he was concerned about that was traveling through Johnstown that day. Passenger manifests for train travel were rarely kept at that time. If you had a ticket to ride you got a seat. The reply to the President read as follows:

Presdt Harrison

Washington, DC

All passengers on trains 8 and 12 that were delayed at Conemaugh were forwarded by wagon to Ebensburg…there were a few drowned but it is impossible to give the names… they lost their lives by not remaining in the cars…had they done so all would have been saved…this in answer to your message to Colonel Hastings…All passengers in Pullman cars were saved.

The telegram reply to the President was not accurate. It stated that the rail passengers who died "…lost their lives by not remaining in the cars." But as noted earlier in this chapter, some of the people who stayed aboard had also died, and some who had the farthest to run survived because they had been warned. The assistant to Hastings who replied to the President did not know that the railroad agent at East Conemaugh, E. R. Stewart, made a list of passengers who survived (PRR testimony, Michael Trump). Stewart sent the list to Ebensburg to help arrange for their travel there and care on arrival, along with revised tickets east. Meantime the railroad people arranged temporary shelter in East Conemaugh for the remaining passengers.

As there were no passenger manifests of who had boarded the trains it remains uncertain how many riders died, but 37 names are in the official list. The dead included Jennie Woolf of Chambersburg, PA, Kate Minich from Ohio, Mrs. Christman from Dallas, Mrs. Brady from Chicago, Mrs. Rainey from Kalamazoo, and many others. The train schedule and fate itself had placed them in the wrong place at the wrong time, in the crosshairs of the flood wave. But Alfred Brown of Chicago and Mrs. Blaisdell of Pelican Rapids were two of the passengers who lived.

5.6   Stories of Other Survivors

David McCullough (1968) presents an extraordinary narrative of the experiences of various survivors in East Conemaugh, in Woodvale, and in Johnstown and its neighboring boroughs. An example was the large group of people with Reverend Beale. McCullough artfully and humanely captures the deprivations, horrors, and also the bravery through the long night that many spent in wrecked or floating houses and roofs, including those trapped in debris piled against the stone bridge below town. That enormous debris jam backed up water over a large area of Johnstown, as the Little Conemaugh and Stoney Creek Rivers were still at flood levels, pouring water into the valley. Eventually part of the railroad embankment on the north side of the stone bridge eroded away and the water at last had an outlet. The somber, debris-filled, backed-up water body slowly drained.

McCullough researched the Johnstown flood while a few survivors were still living. Notable among those was Victor Heiser, who apparently survived because his father had sent him to do a chore in the barn just as the flood wave approached Johnstown. His family lived at 224 Washington Street, the pre-flood address. He narrowly escaped but his family and virtually all possessions were lost. Even his shoes were gone. Heiser was cared for by a local family, received an excellent education, and became a medical doctor. He wrote a book (Heiser 1936) that describes his extensive world travels and professional work. A copy of his now rare work is maintained by the Johnstown Area Heritage Association.

5.7   Flood Heights of the "Great Wave"

After the flood the PRR sent a team of engineers led by Antes Snyder[8] to survey the hillsides along the flood path from the dam down to Johnstown. They mapped the height of the flood wave above the normal river level, fixing the wave heights based on erosion and rafted debris deposited high on the slopes. The greatest wave height they found was ~66–67 feet, just downstream from the Conemaugh viaduct. The next chapter reviews events at the viaduct. Some stretches of railway in the Conemaugh Valley survived, their embankments perched on slopes higher than the local flood crest. For example, from below Mineral Point upstream to the viaduct the tracks were safely above the flood wave and thus preserved. In other areas the rails, tracks, telegraph poles, and parts of embankments themselves were swept away.

Table 5.1 shows a sample of the heights of the water above the normal level of the Conemaugh River at different locations (Snyder et al. 1889). The variations in flood height were mostly caused by changes in the width of the valley along the flood route and also by the temporary formation of debris dams at the viaduct and at Bridge #6. Water backed up behind these "dams" until they were dislodged by the pressure of the rising water. Flood depths were affected to a lesser extent by meanders in the river.

After the flood swept through East Conemaugh, Franklin, and Woodvale the destructive wave emerged from the narrow gorge east of Johnstown and spread in an arc over the broad floodplain that Johnstown and other boroughs had been built on. The height of the water "wave" varied from place to place in the valley, depending on the height of the ground, the amount of entrained debris, and the sturdiness of the buildings encountered. Most structures were obliterated but some remained standing, like the Methodist Church on Franklin Street. The sturdy Alma Hall building also survived and sheltered many survivors overnight. It exists today in Johnstown.

---

[8] Antes Snyder was originally from Lycoming County, PA. He worked for the Pennsylvania Railroad for most of his life, and was so employed at his death. For a time he served as an assistant superintendent under Robert Pitcairn.

Table 5.1 Flood heights from dam and along Conemaugh Valley

Location	Flood height above normal river level
South Fork reservoir (lake depth at dam)	73 ft
River reach immediately below dam	39 ft
South Fork village	33–35 ft
Start of river loop, east end of Viaduct	35–36 ft
At Viaduct	67–68 ft
Mineral Point	31–33 ft
Midway from Mineral Pt. to Bridge #6	32–36 ft
Bridge #6 (short distance *above* bridge)	43 ft
Bridge #6 (short distance *below* bridge)	43 ft. (valley very narrow)
Buttermilk Falls	43 ft
Conemaugh Station at East Conemaugh	29 ft
Woodvale	24 ft. (less depth due to valley widening)
Johnstown City Hall	21 ft. (depth of spreading tsunami-like wave)

The building that served as city hall in 1889 was destroyed by the flood. A new structure crafted with massive walls was built in 1900. Its walls withstood the terrible floods of 1936 and 1977. Flood levels from those years and 1889 are shown as plaques on the corner of City Hall at the intersection of Main and Market Streets (Fig. 5.2). After the 1936 flood the Army Corps of Engineers did extensive work on the river channels to try to prevent or greatly reduce damage from future floods. But the 1977 water level at City Hall shows that great floods could still inundate the lower streets of Johnstown.

Johnstown and its adjacent boroughs were destroyed by a powerful, tsunami-like wave. In the streets the debris was pushed and tumbled as much as carried along by the flood. There exist at YouTube.com extraordinary videos of tsunami destruction. The massive Tōhoku earthquake in Japan, in March 2011, spawned tsunami waves of tremendous power. Some waves reached heights of 133 feet in Miyako in Iwate Prefecture. In 2004 an earthquake greater than magnitude 9 struck off the west coast of Sumatra. It produced tsunami waves as high as 100 feet and killed hundreds of thousands of people.

The ways in which tsunamis destroy a coastal town can be seen in the videos captured by terror stricken residents. Seeing those videos helps us imagine what the people of Johnstown and its neighboring boroughs saw with their own eyes that terrible day in 1889. The debris-filled wave poured and tumbled down the city streets. Some survivors in Johnstown and elsewhere reported seeing buildings rising up or being torn apart with no obvious cause. Rather than seeing water, they witnessed houses ripped apart from beneath, the debris then pushed along and rolled under in a grinding bow wave of destruction.

Unless shielded in some way or on the margins of the flood, human beings in the path of the Johnstown flood had little chance to survive. Some, like Victor Heiser, were very lucky to catch a precarious ride on floating debris that was not driven under the water.

Fig. 5.2 Flood levels from
1889, 1936, and 1977
marked with plaques at
corner of Johnstown City
Hall. Photo 2017 by the
author

References

Barton CH (1910) The Red Cross – in peace and war. American Historical Press, p 703

Brown AH (1889) Letter: history of the flood of the Conemaugh Valley, as witnessed by Alfred H. Brown, May 31, 1889, 6 pages (letter written from 44 Tuers Avenue, Jersey City). Copy of letter in archive of the Johnstown Area Heritage Association

Connelly F, Jenks GC (1889) Official history of the Johnstown Flood, Illustrated. Journalist Publishing Co., Pittsburgh PA, p 252

Degen P, Degen C (1984) The Johnstown Flood of 1889: the Tragedy of the Conemaugh. Eastern Acorn Press, p 64

Eng. News (1889a) Engineering News, Jun 22 1889, p 563

Eng. News (1889b) Engineering News, Jun 15 1889, pp 551–553

Gazette (1889) Pittsburgh Commercial Gazette, Jun 4th

Heiser V (1936) An American doctor's odyssey – adventures in 45 countries. WW Norton & Co., Inc, New York, p 535

Kaktins U, Davis Todd C, Wojno S, Coleman NM (2013) Revisiting the timing and events leading to and causing the Johnstown flood of 1889. PA Hist A J Mid-Atlantic Stud 80(3):335–363

McCullough DG (1968) The Johnstown flood. Simon & Schuster, New York, p 302

McLaurin JJ (1890) The story of Johnstown: its early settlement, rise and progress, industrial growth, and appalling flood on May 31st, 1889. James M. Place, Publisher, Harrisburg

PRR (1889) Statements of employees of the Pennsylvania Railroad Company, and others in reference to the disaster to the passenger trains at Johnstown, taken by John H Hampton, at his office in Pittsburgh, by request of Superintendent Robert Pitcairn; beginning Jul 15 1889

PRR (1892) Pennsylvania Railroad Company. In Memorium: John Henry Hampton. Jos. Eichbaum
    & Co., Pittsburgh, p 26
Snyder A, Vandivort T, Wakefield AE (1889) Map of the valleys of the South Fork and [Little]
    Conemaugh Rivers, showing the path of the Great Wave from its origin in the Western Reservoir
    to and beyond the Stone Bridge at Johnstown, Cambria County, PA
Tribune (1889) The Bursting of the Dam. Johnstown Daily Tribune, Jun 25 1889, XVII(5087):1

Chapter 6
The Viaduct

About 2.2 km west of South Fork, the tracks of the Pennsylvania Railroad (PRR) crossed the Little Conemaugh River on a high bridge, spanning a gorge where the river bent back on itself like a coiled snake. Known as the Conemaugh Viaduct, the imposing but beautiful single-arch bridge was built as part of the Portage Railroad in 1832–33 (Fig. 6.1). The dressed stones in the viaduct were obtained from nearby surface boulders and bedrock outcrops, including sandstones and limestones. No mortar was used for much of the main structure, the courses of stones being placed one atop the other and well fitted. Dressed stones were used for facing the bridge and the slope wall on the western side. These were placed in a mortar made from the Loyalhanna[1] Limestone that crops out along the gorge.

Civil engineer Solomon Roberts wrote a description of the Conemaugh viaduct in 1878. He had trekked with the original survey party that sited rail lines and bridges. He also supervised the eventual building of the viaduct. Photographs from the post-Civil War period show picnickers in Sunday finery posing in the gorge along the river with the bridge and its arch in the background. Originally designed with a double arch, Chief Engineer Welsh later settled on a plan with a single graceful arch that spanned 80 feet. Roberts wrote that the top of the viaduct was about 70 feet above the valley floor. The hillside at the southeastern end of the bridge is very steep, and here the builders, John Durno[2] and his gang of stone masons, fitted wingwall stones directly into notches cut into the bedrock. A small part of the original southeast wingwall remains on the hillside today, sited on private property (Fig. 6.2).

A large earthen embankment was built on the northwestern side of the gorge to carry the tracks from the viaduct to the mountainside. The embankment itself reached a height of 64 ft. Detailed sketches published by *Eng. News* (1889) show the eroded end of the embankment and its masonry slope wall, a small part of the

---

[1] The Loyalhanna is a distinctly cross-bedded limestone of Mississippian age, >323 million years.

[2] It is unfortunate that the viaduct did not survive as a monument to Durno's skill. He later died after falling from another high bridge.

© Springer International Publishing AG, part of Springer Nature 2019
N. M. Coleman, *Johnstown's Flood of 1889*,
https://doi.org/10.1007/978-3-319-95216-1_6

Fig. 6.1 Illustration of the Conemaugh viaduct from upstream (from Roberts 1878)

arch foundation on the western side, and fragments of the arch foundation and a wingwall on the southeastern flank of the gorge (Fig. 6.3).

On the day of the flood, after reaching the village of South Fork the flood wave struck the hillside beyond the PRR tracks and momentarily divided, part of the deep flow driving a reverse flow to the east, up the Little Conemaugh River. The bulk of the flow sped westward down the main channel of the river, and this was soon joined by backwash from the reverse flow. Before reaching the Conemaugh Viaduct the river channel turns sharply south. Most of the water went that way, coursing round the 2.6 km-long river loop to the viaduct. But the floodwater was deep, and as it struck the hillside at the sharp river bend the wave struck the hillside and pushed a cascade of water through the railway cut in the narrow neck of land at the eastern end of the viaduct, forming a waterfall of muddy water and debris. The railway "cut" (see Fig. 6.4) had been excavated by railway engineers to carry the tracks on a level through the narrow neck of land where the river gorge looped back on itself. Wave-height mapping by Snyder et al. (1889) showed that floodwater ~15 ft. deep was driven through the cut at the flood's peak. Erosion from the resulting "waterfall" may have damaged the arch foundation on the eastern side and contributed to the eventual collapse of the viaduct.

Fig. 6.2 Wingwall remnant of the original viaduct. These dressed stones were keyed into the bedrock along the side of the gorge, the blocks carefully shaped and placed atop each other. After the flood the PRR erected a temporary wooden trestle to restore rail service. It was completed on June 12, just 2 weeks after the flood. The stone viaduct itself was rebuilt in 1889, part of which appears in sunlight at upper left. Photo 2015 by the author

As the bulk of the flood wave charged round the river loop it gathered more debris, adding to the mass of material torn up and carried from South Fork and the rail embankments. The debris included rail ties, parts of buildings from the low parts of South Fork,[3] trees, bushes, and idle railroad equipment. In the steep gorge of the river loop the flood tore loose virgin timber; hemlock, oak, beech, and chestnut trees of enormous length and girth. The virgin trees in the region were being logged off or burned, but not those in the steep, hard-to-access river gorges. Many of these trees were over 100 feet tall, and a few of them lodged across the 80 foot viaduct arch, with their enormous spreading tops, would have formed an effective

---

[3] Many buildings in South Fork were spared because they were built on a hillslope higher than the flood wave could reach.

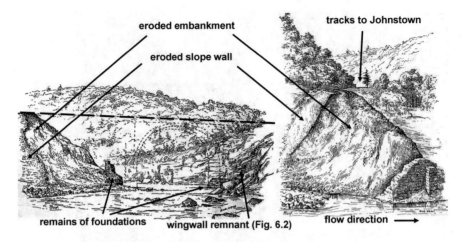

Fig. 6.3 Sketches of viaduct remnants by *Engineering News* staff member Frederic Burt, (Frederic Percy Burt was born Feb. 3, 1861 in Warwickshire, England. His parents were Arthur and Anne Jones (Bird) Burt. He came to the U.S. in 1871; was educated in Canada and from 1882–86 worked there in civil engineering and surveying. Had long association and held various positions with Engineering News, through 1904, including vice president and manager. Later was president and director of The American Architect (Leonard 1907)) who visited immediately after the 1889 flood and before a wooden trestle was built (*Eng. News* 1889 p 570)

Fig. 6.4 *Left*: Post-flood photo from June, 1889 showing height of flood-transported debris and stripped and abraded standing trees. The temporary wooden trestle was nearing completion. Image courtesy of the Johnstown Flood Museum Archives, Johnstown Area Heritage Association [original photo by R. K. Bonine (d. Sep 11, 1923)]. *Right*: LiDAR map showing extent of temporary lake that formed in river loop above viaduct. Circled arrow at top shows where photographer stood and pointed to capture photo in left frame [Credit: S. Wojno]

sieve to capture smaller trees and bushes. The result must have looked like part of a monstrous beaver dam made of whole trees. The obstruction created an effective "dam" that briefly slowed the flood wave, accumulating water in the river loop. The long earthen embankment supporting the rails west of the viaduct formed the longest part of this "dam." The narrow river gorge grew wider at the bridge location, but that long embankment across the valley forced all the floodwater to funnel through the viaduct arch. But the partly obstructed arch caused water to rapidly accumulate in the river loop. Using LiDAR and assuming a water depth reaching 67 ft., the "lake" volume in the meander loop was ~3 × 10^6 m^3, or a little less than one-fifth of the volume in Lake Conemaugh at the time the dam breached.

The viaduct stood long enough for the floodwaters to deeply fill the river gorge. The evidence for this was three-fold. First, the height of the flood wave mapped by Snyder (1889; see previous chapter) showed erosion and debris deposition up to ~66–67 ft. above the "ordinary stage" of the river. Second, post-flood photographs show massive flood-rafted logs deposited high above the river bottom on the northwest side of the gorge, where bark and branches were also stripped from standing trees (Fig. 6.4). And third, on the northwest side of the viaduct, the earthen embankment and its slope wall were eroded by deep floodwaters that surged through the gap created when the viaduct collapsed (Fig. 6.3).

Given the size and massive weight of the stone structure (28 feet wide at the top and 40 feet wide at its base), the viaduct and its obstructed arch briefly stood the pressure of the accumulated "lake." Finally, the unreinforced layers of dressed stones gave way and the viaduct arch and its supports collapsed, releasing the debris dam and a renewed flood wave down the valley. There was an old rumor, certainly false, that a dynamite shed had been borne along in the flood until it struck the viaduct and exploded, destroying the bridge. Such rumors likely came from people who did not understand the power of deep floods, who thought the viaduct was too strong to be felled by mere water. The only wonder was that the viaduct stood as long as it did. But the sudden collapse would certainly have made a tremendous crashing or booming sound, not unlike an explosion.

6.1   River Flow from Collapse of the Viaduct

In 2009 I performed hydraulic calculations based on dam breach theory for the discharge following collapse of the viaduct. The discovery of the *Engineering News* sketches in Fig. 6.3 makes it possible to repeat those calculations with updated breach dimensions. The sketches show the size of the breach with remnants of the old viaduct still in place on both sides of the river. Analytical solutions of breach flow are typically based on a symmetrical breach of trapezoidal or rectangular shape, with sides of similar slope. It is first necessary to define the shape and dimensions of the breach; its depth, bottom width, and side slopes. The present-day twin arch viaduct is not helpful for this because its design differs from the single arch bridge built in 1833. What both had in common was being attached to a bedrock slope on the southeastern side, with an earthen embankment on the other side. The original viaduct had vertical, curving wingwalls flanking both ends of the bridge. The replacement bridge has no wingwalls on the southeastern end, the stonework being fitted perpendicularly into the bedrock hillside. On the northwestern end there is a vertical, straight wingwall perpendicular to the railway, but only on the downstream side of the bridge.

The geometry of the opening that was created when the viaduct collapsed can be estimated using nineteenth century information about the viaduct, including pre-flood photographs and the post-flood sketches by Burt (Figs. 6.3 & 6.4), plus modern LiDAR data. The eastern side of the gorge was very steep with exposed bedrock. The left frame of Fig. 6.4 reveals in the distance bedrock exposed on the southeastern slope after floodwater stripped away the vegetation and thin soil. That hillside was therefore robust and its slope unlikely to have been changed by the flood or by erosion in the intervening years. An inclinometer yields a range of slopes from 40–45 degrees for the eastern side of the valley. On the northwest side of the river, sketches made in June 1889 (Fig. 6.3) show that the end of the embankment had an eroded, steep slope toward the water. The shearing away of the vertical stone wall at that side of the river destabilized and carried away the upper part of the adjacent embankment. Abrasion from transported debris would also have contributed to the eroded slope seen in the post-flood sketches and photographs. The embankment slope was rebuilt and cannot now be measured in the same way as the bedrock slope on the eastern side, but it was very steep. I used the same range of slopes for both sides of the breach: 40–45 degrees.

The pre-flood photographs and preservation of viaduct remnants help to define the bottom width of the flow section. The sketches in Fig. 6.3 reveal these viaduct remnants and reflect key observations by civil engineer Frederic Burt. On the southeast side a part of the arch base was preserved, and on the northwest was a remnant of the vertical stone wall that abutted the earthen embankment. The foundation stones for the western side of the arch do not appear in the sketches. They were either torn out by flood erosion or were preserved in shallow water for Burt's observation. Using the 80 ft. (24 m) wide arch as a scale provides a bottom width for the flow section in the range of 33 to 36 meters.

To estimate the discharge at the time the viaduct collapsed it is necessary to estimate the maximum water depth at that time. Roberts (1878) wrote that the viaduct was about 70 feet high. The top of the present viaduct is about 85 ft. above the river at an elevation of 1463 ft. (446 m). It is tempting to use the full height of the original viaduct to calculate the discharge, but there is evidence the water was not so deep. As shown in Table 5.1, Snyder et al. (1889) in mapping heights reached by the "Great Wave" recorded a flood level of 67–68 ft. on the hillside downstream from the viaduct. Assuming an "ordinary" river depth of 2 to 3 feet I estimate that the temporary lake upstream of the viaduct reached a depth of ~70 ft. (21 m) just before the structure collapsed.

Equation 11b of Walder and O'Connor (1997) was used to estimate the maximum flow through the breach. I review this equation and its variables in Chap. 10 of the present work. The maximum discharge rate would have been in the range from

7465 to 8484 m^3 s^{-1}, using the range of geometry for the flow section. These discharges should be viewed as an upper limit because the analytical approach assumes water flow. In fact the initial flow through the opening was debris-choked water. With the bridge gone the densest part of the debris jam should have cleared rapidly so that in a minute or less most of the flow would have been water. Some lowering of the water level in the river loop would have happened in that time, therefore I reduce the discharge estimates by 5% to obtain a peak discharge range of ~7090 to 8060 m^3 s^{-1} following collapse of the viaduct and clearing of the debris jam. As will be shown in subsequent calculations, this is similar to the lower range of discharge through the breach of the South Fork Dam. In effect, the temporary blockage at the viaduct reinvigorated the power of the flood wave. Discharges would have rapidly declined from the peak flow rate following the viaduct collapse.

6.2 The Viaduct Today

You can see the present-day viaduct by visiting the village of South Fork. There is a wide and pleasant walking path on the north side of the Little Conemaugh River, known as the "Path of the Flood Trail." The trail is accessed from a small parking lot at the western end of Portage Street, on the north side of the river. The viaduct can be seen from several overlooks on the trail, but the best views happen from late fall through early spring when the trees and shrubs have lost their leaves (Fig. 6.5).

The river loop where water backed up behind the Conemaugh viaduct is visible from various hilltops, including the farm on high ground where Uldis Kaktins lived. For 40 years he and his family lived there and witnessed the river valley in every kind of changing weather.

In 2010 I visited Professor Kaktins' home with students from my university class. He had offered the use of the property for geophysical surveys in a field project. I arrived early before the students on a sunny autumn morning and parked below the house. As I climbed out of the car Uldis came down the back steps with two steaming cups of coffee and handed me one.

"Look at that" he said, pointing north. "You don't see that every day." I did not have a camera along, or the reader would see that image here.

From Kaktins' high hilltop we looked over the broad drainage basin of the Little Conemaugh River. A layer of fog had formed in the bottom of the river valley on that windless morning. This was advection fog, produced in the shaded valley over the cold river water. There was the river loop through which the floodwaters charged in 1889. The fog filling the deep gorge was dense with a well-defined, flat top that looked just like a deep lake in the river curve. This view was how the river looked on the day of the flood (Fig. 6.4, right frame), except on that terrible day a real lake had gathered in the coil of the river, soon to be released as a renewed flood wave when the viaduct collapsed. The villages of Mineral Point and East Conemaugh and its train yard and brick roundhouse were in the direct path. Between those villages were bridge #6 and another large loop in the river, just northeast of the present

Fig. 6.5 Professor Emeritus Uldis Kaktins at an overlook, viewing the present-day Conemaugh Viaduct. This twin-arch bridge was built here in 1889 to replace the destroyed single-arch structure. The original viaduct was similar in height and length but much narrower on top, < 30 ft. Photo (2012) by the author

village of Parkhill. That river loop no longer exists, having been filled in with waste material largely from the steel-making industries. Just beyond bridge #6 was the AO tower where Michael Trump and his railroad men were standing on the afternoon of the flood. The flood wave then struck East Conemaugh and cruelly descended on the towns farther below.

The cause of it all was the modification of the South Fork dam by the South Fork Fishing & Hunting Club. In particular, their lowering the dam crest critically reduced the discharge capacity of the main spillway and eliminated the action of an auxiliary spillway. The Club's actions made the dam susceptible to overtopping during great storms like that of May 30–31, 1889. The original dam built by the State would have allowed the floodwaters to safely pass (see Chaps. 9, 10, and 11).

In the wake of the Johnstown Flood, the American Society of Civil Engineers was about to hold their regular meeting in New York. They discussed the disaster, clearly due to the failure of an engineered structure, and appointed a committee of prominent engineers to travel to South Fork to investigate the cause of the dam breach. And there our story continues.

References

Eng. News (1889) "The work of the flood at Johnstown," Engineering News, Jun 22 1889 p 570

Leonard JW (1907) Who's who in New York city and state, LR Hamersly & Co, NY, p 223

Roberts SW (1878) Reminiscences of the first railroad over the Allegheny Mountain. In: Pennsylvania magazine of history and biography, vol 2. The Historical Society of PA, Philadelphia, pp 370–393

Snyder A, Vandivort T, Wakefield AE (1889) Map of the valleys of the south fork and [little] Conemaugh rivers, showing the path of the great wave from its origin in the western reservoir to and beyond the stone bridge at Johnstown, Cambria County, PA

Walder JS, O'Connor JE (1997) Methods for predicting peak discharge of floods caused by failure of natural and constructed earthen dams. Water Resour Res 33(10):2337–2348

Chapter 7
Investigation by the ASCE Committee (Francis et al. 1891) and the Fate of their Report

7.1 ASCE's Investigation and Sealing of the Report

The American Society of Civil Engineers (ASCE) held its regular meeting in New York on June 5, 1889, just five days after the Johnstown disaster. Vice President Alphonse Fteley chaired the gathering. Meeting notes appeared in the June 8th issue of *Eng. News* (1889a, 532–533).

The Secretary announced the deaths of two members, one of whom died at Johnstown. He was 34-year-old Charles A. Marshall, engineer of tests for the Cambria Iron Company. On June 4th the *Courier-Journal* newspaper in Louisville, KY reported that "No tidings were heard yesterday from Mr. Charles Marshall, formerly of this city." He was one of the 48 lodgers who died at the ill-fated Hulbert House[1] on Clinton Street in Johnstown, and this terrible news soon reached his wife. His body was later identified and transported to Louisville.

Chairman Fteley discussed the dam breach at South Fork. In 1874 the ASCE had sent a committee to investigate the fatal breach of the Mill River dam in Connecticut. Fteley said it would be proper to pursue the same course now. William Shinn of Pittsburgh gave a brief summary about the dam and reservoir based on early reports of the breaking of the dam. Some discussion ensued, and a committee of four was appointed to investigate the cause of the Johnstown disaster: Mr. M. J. Becker, Mr. A. Fteley, Mr. J. B. Francis, and Mr. W. Worthen. Fteley informed the gathering he had also been directed by the New York Aqueduct Commissioners to study the case and write a report to them.

In the aftermath of the Johnstown flood this prestigious committee was to visit the South Fork dam as soon as rail travel was restored, and to investigate the cause of its failure. Max J. Becker was ASCE's President in 1889, but had virtually no experience with dams or canals. Francis was a hydraulic engineer best known for

---

[1] The *Johnstown Tribune* of July 5, 1889 gave the names of 12 lodgers at the Hulbert House who were saved.

© Springer International Publishing AG, part of Springer Nature 2019
N. M. Coleman, *Johnstown's Flood of 1889*,
https://doi.org/10.1007/978-3-319-95216-1_7

his work in flood control, turbine design, canal work, dam construction, and weir discharge calculations. He was a founding member of the ASCE and served as its president from November, 1880 to January, 1882. He was an emeritus member, mostly retired by 1889, and though still involved in consulting work he had turned over most of his responsibilities in Lowell, Massachusetts to his son, James. Alphonse Fteley was one of the two Vice Presidents elected in 1889, and was a protégé of William E. Worthen. Worthen was a past President of the ASCE and well known for his extensive work in New York and New England. Becker then resided in Allegheny City, now known as Pittsburgh's North Side, while Fteley and Worthen lived in New York.

A fifth expert was nominated but declined to join the investigation. An item in *Eng. News* of June 15th (1889b) explained that William J. McAlpine "...would have been a very desirable fifth member, if he would have consented to serve. Probably no man, in this country, living or dead, has had a larger experience in the construction of dams of this class." McAlpine was also a former president of the Society and previously served as New York State Engineer and Surveyor. But he died on February 16, 1890, less than 9 months after the Johnstown flood. It is possible he had health problems in 1889 that made him reluctant or unable to participate and knew he could not give his best skills to this inquiry.

Due to the "...present excitement and interest in regard to dams," the June 15th issue of *Eng. News* (1889b) also contained an article about large water-supply dams in the U.S, those 40 feet high or greater. The authors counted 34 such dams, not including many other sizeable dams built to support canals, irrigation, and manufacturing. Seven other dams were included between 34 and 40 feet high. A number of water supply dams exceeded the height of the South Fork dam: a dam at Cambridge, Mass. (85 ft); Boyd's Corners Dam (78 ft) and Middle Branch Dam (91 ft) in New York; the Eden Park Dam in Ohio (119 ft); Sweet Water Dam in California (98 ft) and a dam near Oakland (80 ft); the Pilarcitos Creek dam near San Francisco (95 ft), and a "Dam on San Andreas Supply" with a height of 93 ft. By 1889 the South Fork dam was just one of the larger dams in the country.

Railway access to the Johnstown region was reestablished by June 13th, open from Pittsburgh to Altoona. The Committee visited the South Fork dam site and downstream locations, reviewed the original design of the dam and subsequent modifications made during repairs, conducted an elevation survey of the dam remnants, interviewed neighbors and eyewitnesses, and performed various hydrologic calculations.

At first it was difficult to learn who truly chaired the ASCE investigation committee. Francis' name appears first in the list of authors in the report submitted to the Society on Jan. 15, 1890. In that list (see excerpt below from Francis et al. 1891 p 457), James Francis heads the list with his first name given, which suggests he was chairman. Max Becker was the standing president of ASCE and a railroad man. His name was listed third on the final report, but as it came out later, he was the de facto chairman of the investigation. The report was sealed at that time and not given to the Society's members, even though many clamored for it. Becker decided not to release the report, and his fellow committee members were displeased about that. At that time they deferred to him as president. They could not have known then that Becker and two other presidents of ASCE would keep the report under wraps for another year and a half.

There are to-day in existence many such dams which are not better, nor even as well provided with wasting channels as was the Conemaugh Dam, and which would be destroyed if placed under similar conditions. The fate of the latter shows that, however remote the chance of an excessive flood may be, the only consistent policy, when human lives, or even when large interests are at stake, is to provide wasting channels of sufficient proportion and to build the embankment of ample height.

Respectfully submitted,
James B. Francis,
W. E. Worthen,
M. J. Becker,
A. Fteley.
January 15, 1890

In the June 15, 1889 issue of *Eng. News* the following appears. Becker's name is listed first, though his first name and initial were switched. It is curious why an engineer with virtually no experience in dams or hydraulic engineering would be on the committee, let alone assume chairmanship of an investigation of a dam disaster that killed thousands of people.

A COMMITTEE TO INVESTIGATE the Johnstown disaster, consisting of Messrs. J. MAX. BECKER, President, JAS. B. FRANCIS and WM. E. WORTHEN, ex-Presidents, and A. FTELEY, Chief Engineer of the new Croton Aqueduct, has been appointed by the American Society of Civil Engineers, and will proceed to the scene of the disaster as soon as railway communication is re-opened....

All four men attended the Society's annual convention at the Octagon House in Seabright, NJ during June 20–25 where Becker as President gave the annual address (ENARJ Jun 29 1889). On the afternoon of Saturday, June 22, Vice President Fteley gave a preliminary report about Johnstown, noting there was much information yet to be gathered. He felt it premature to state causes about the dam break given that the Society's conclusions would bear considerable weight. However, conversely, he then went on to say there was no doubt the disaster would not have happened if the spillway had adequate capacity.

The committee had just returned from visiting the South Fork dam and travelling through the flood-ravaged central valleys of Pennsylvania. They also covered the flood route from South Fork to Johnstown, witnessing the destruction wreaked by the flood wave. Fteley declared it an interesting sight to see three officers of the Society, the President and two past presidents, carrying and using the survey field equipment. His comments confirm that the committee made survey measurements themselves, clambering over the physical remains of the South Fork dam. They were focused on key questions about the spillway capacity, the volume of the reservoir, and the size of the watershed. A complete survey would be commissioned for the watershed and flood-inundated margin of the former lake. Fteley said:

> "It is very essential that the truth of this matter should be known, as it may show that other dams are unsafe in the light of recent experience." [*Eng. News*, 6/29/1889d, p 603]

Clearly, Fteley's remarks before the membership on June 22nd reveal that safety about other dams was, in his view, the driving force behind the investigation. It is therefore astonishing that the Committee's final report would be completed in 7 months but then remain sealed until 2 years after the disaster!

Max Becker gave the President's plenary address at the convention in New Jersey. An extended abstract of his talk appears in the June 29th issue of *Eng. News* (1889c). He spoke on a wide array of topics, including the promise of electrical engineering, rapid transit (a term in use even then), electric railways, water works, sanitary engineering, the need for improved streets and highways, canals and hydraulic engineering, and an extended discourse in his main area of expertise, railroads.

Of the Johnstown disaster President Becker said that the cause of the failure of the South Fork dam is being investigated by a committee appointed by the society, which had just returned from visiting the scene. He went on to say that:

> Examinations and measurements of the structure and its surroundings, and extensive information obtained from various sources, will enable the committee to submit to the society in due time a comprehensive statement of the conditions and circumstances which have induced and contributed to this most disastrous failure.

Becker also related the experience from Johnstown to the massive dams used to construct reservoirs to supply water to large cities:

> The importance of constructing these dams of proper shape and size, and of suitable material and good workmanship, so as to insure their absolute strength, and give them sufficient resisting capacity against every possible contingency, has been taught by a recent lesson of frightful experience; and while the responsibility for this calamity may not be placed upon the shoulders of the profession, yet it will be well for its members to look upon it and remember it as a warning and an example.

An additional report about Johnstown was planned for the members at the first meeting in September. A summary of the September 4th meeting in *Eng. News* (1889e) made no reference to Johnstown. However, the September 14 issue (1889f) gave the results of the commissioned survey of the watershed and lake area at South Fork and credited the work to Mr. A. Y. Lee, a civil engineer from Pittsburgh. The article stated (p 259):

> The main fact we can be sure of is that, with a mountain watershed of this area and a reservoir of this size, the original spillway provided for the dam was nearly, if not quite, enough for perfect safety, while the obstructed spillway, as it was when the dam broke, was less than half large enough.

The committee finished their investigation of the South Fork dam breach and submitted the report on January 15, 1890 at the annual meeting of the Society at its headquarters in New York. Elections were held at these annual meetings, and

Alphonse Fteley was reelected to serve as one of the two Vice Presidents. The new President to replace Becker was William P. Shinn. The South Fork investigation was briefly discussed. Engineer Don Juan Whittemore objected to the appointment of a committee on standard rail sections. He preferred to leave that to railway managers and the engineers of rolling mills, and he considered the work of "…such committees as this and as the one on the Causes of the South Fork Dam Disaster was beyond the purpose of the Society [ASCE]." That was an extraordinary statement from an engineer and Doctor of Laws (Univ. of Wisconsin) on the heels of a dam breach that had swept more than 2200 people to their deaths. Whittemore did not clarify his perceived "purpose" for the ASCE. He was Chief Engineer for the Chicago, Milwaukee & St. Paul Railway, a prominent railroad man, objecting to an investigation properly in the domain of hydraulic engineering. In Whittemore's railroad position he would certainly have known and interacted with Max Becker.

The following text about South Fork from the January meeting appeared in *Eng. News* (1890a, p. 58):

> The committee on the South Fork Dam reported progress, stating that in view of pending suits against the owners of the dam, it was not deemed expedient at this time to publish the full report. This full report will be signed and sealed and entrusted to the chairman of the committee until such time as it may seem proper to publish it.

So there it was, the note in *Eng. News* telling the engineering fraternity and the world that ASCE's investigation of the South Fork dam was complete but was now sealed, to be released at some future time. And although the name of James Francis headed the signature page, Max Becker was the *de facto* chairman of the committee. Becker's last action as President was to seal the report and not release it. Even though he was no longer president, he was being permitted to retain control – to decide when to let it out. And that was apparently OK with ASCE's new president, William Shinn, a railroad and steel man who once had been a managing partner of Andrew Carnegie's. Shinn could have released the report immediately on his authority as president, but he kept it sealed for his entire term.

7.2   Annual Convention in Cresson, PA, June 1890

ASCE's annual convention in 1890 was to be held in Cresson, Pennsylvania, just 12 miles east of South Fork via the Pennsylvania Railroad. The meeting place was "…a large hotel managed under the direction of the railroad company." (*Eng. News* Apr. 19, 1890b, p. 371) The cost for attendees staying at this PRR hotel was reduced to $3/day. A special train was reserved for the convention attendees from New York. That train left the city Wednesday morning, June 25, at 10 a.m., running as a second section of the Chicago Limited. The train would arrive at Cresson in the afternoon.

A notice in the June 7th issue of *Eng. News* (1890c) gave the titles of nine new papers to be presented at the Cresson convention, along with reports by named special committees (revision of the constitution, uniform standard time, compressive strength of cements and the compression of mortars and settlement of masonry, uniform methods of testing materials, and standard rail sections). Other papers were announced on June 14th bringing the total to 15. But the South Fork investigation was not among the reports to be given. Apparently the time was not yet "right." That same issue of *Eng. News* rejoiced in the fact that "…from their titles they are excellent papers. This list is especially gratifying as showing that this Society is, after all, not so poverty stricken in the line of papers as the baldness of the notices of meetings for some time past would quite naturally lead an observer to suppose." Certainly the long awaited report on the South Fork dam would have been safety-focused and far more interesting to the members than "One Way of Obtaining Brine," which was a planned paper. A final meeting notice was posted in *Eng. News* (June 21, 1890d, p. 586). It read in part:

> The general arrangements for the Convention are as have been previously announced. The details of the programme are in charge of a local committee at Cresson, composed of members of the society and officers of the Pennsylvania R.R. residing in that vicinity. These details cannot yet be fully announced. It is expected, however, that the Convention will begin its sessions Thursday morning, June 26. The President's address will probably be made Thursday evening. Thursday, Friday, Saturday and Monday will be occupied partially by meetings and partially by short excursions by rail to points not far from Cresson, probably including Altoona, Johnstown (Cambria Iron Works), Bell's Gap. The banquet will occur, as now arranged, on Saturday evening.

What is curious about the above notice is that the convention program details were not available even 5 days before the start. This was the first annual meeting of ASCE after completion and sealing of the South Fork investigation report, and surely the proximity of the meeting to South Fork merited an excursion. But officers of the PRR had taken a special interest in this meeting and were part of the local organizing committee. Those officers included the powerful PRR executive Robert Pitcairn, who was also a member of the South Fork Fishing & Hunting Club, and his subordinate manager Michael Trump. And contrary to the statement in the meeting notice, they were from Pittsburgh and did not reside in the vicinity. The convention program excluded any visit to the South Fork dam.

The Cresson convention opened on Thursday, June 26, with 175 members and guests and 61 accompanying ladies. The presidential address was given that evening, followed by a lengthy talk with 300 slides of the Grand Canyon. The images were from a recent trip through the entire canyon, the first since Major Powell's expedition. A condensed version of the Presidential address by William Shinn was published by *Eng. News* (Jul 5 1890e). He briefly mentioned the failure of the South Fork dam, that it "… undoubtedly shows the danger of a too contracted spill-way. A similar cause brought about the failure of the Walnut Grove Dam in Arizona." But what is particularly interesting is another topic raised by Shinn: expert testimony. No doubt that was also much on the minds of Club members at that time. Shinn referred to an opinion previously expressed by William Worthen, about how "unseemly" it was to see two eminent engineers arguing against each other in court. Shinn's concern was that the differences of opinion that "naturally must exist" would lead laymen to lose respect for the profession. Shinn then went on to say,

It has long been my opinion that the expert witness should be called by the court and that he should be allowed to state facts and opinions freed from the trammels of the "*Suppressio Veri*" [a lie of omission] so often employed by the attorneys on either side. It might result in some of my professional brethren receiving less fees, but it would be to the benefit of the profession and the cause of justice.

But I must point out how convenient that would be for powerful companies like the Pennsylvania Railroad, or societies like the South Fork Fishing & Hunting Club, if they could manage to insert court-appointed experts that happened to favor their positions. The courts were often the last resort for those seeking justice from wrongs inflicted by the powerful, and independent expert opinion was often the only way to get a fair result.

Next morning was the excursion from Cresson to Johnstown, which ended with a tour of the town and Cambria Iron Works where Alexander Hamilton was the General Superintendent. The rail excursion included…

…stopping at various points along the Conemaugh River, where the flood had done its most terrible work. The visit was signalized by the discovery the same morning of the bodies of three women who had been buried many feet deep under sand and gravel in the bed of the river, and which were yet, after the lapse of over a year, in a fair state of preservation. Of the work of the flood but little direct evidence now remains, except a general rawness and newness in the general aspect of the town, from which it must take many years to fully recover. [*Eng. News*, July 12, 1890f, p. 34]

About 200 attendees, including many ladies, had disembarked from five coaches in Johnstown just before 10 a.m. They were handed pamphlets and diagrams about the iron industry and maps of the Conemaugh Valley. Everyone climbed aboard the train around noon to return to Cresson, where an evening banquet was held at the Mountain House. An article about the excursion in the *Johnstown Daily Tribune* (6.27.1890) included this note about South Fork:

This afternoon a business meeting was to be held at which it was expected the report of the committee appointed to examine the South Fork dam would be read and considered in secret.

Thanks to the power of the press, we see that the South Fork report had not only been sealed but was now being reviewed "in secret," at a convention where Robert Pitcairn was an organizer. He was not an ASCE member, but clearly had the chance to suppress this engineering investigation related to his private fishing and hunting Club. And William Shinn was president of ASCE. It is no wonder the South Fork investigation report stayed under wraps.

A second excursion took place on Saturday morning, to Bell's Gap. The business session of the meeting was held that afternoon, and that evening a reception for the guests was held in Altoona by Mr. and Mrs. Theodore Ely. Ely was another railroad executive, the Superintendent of Motive Power for the PRR. Many members left the convention after that reception or on Sunday. The final session was on Monday morning. After a last set of "regular" papers was given, and after many attendees had gone home, came a talk about the history of the South Fork Dam by civil engineer Antes Snyder of the PRR. His talk was illustrated "…with maps, profiles and blackboard diagrams. A very animated and interesting discussion followed on the

Fig. 7.1 Portrait of
William Shinn, President
of ASCE in 1890 (graphic
from supplement to *Eng.
News*, Dec 20 1890g)

failure of the dam last year and the destruction in consequence of the Pennsylvania
R. R. viaduct bridge No. 6 in the Conemaugh Valley about two miles below the
dam..." *Eng. News* (July 12 1890f, p 35) went on to say...

> The report of the committee on the South Fork's dam disaster is still withheld. With this
> session the convention closed.

It was at this Cresson meeting that Max Becker, the prominent railroad engineer,
told a reporter that "We will hardly [publish our] report this session, unless pressed
to do so, as we do not want to become involved in any litigation" [*Johnstown Daily
Tribune*, 6/27/1890]. The article, citing material from the Pittsburgh Times, stated
that Becker was the Chairman of the Committee. Becker was no longer President of
ASCE and held no other Society office at that time. But this powerful railroad man,
with virtually no experience in hydraulic engineering, retained iron-clad control of
the release of the South Fork report. President William Shinn could have overruled
him but did not. The report remained sealed.

So who was William Shinn (Fig. 7.1)? Born in Burlington, NJ, he came to
Pittsburgh at age 16 and remained there most of his life, forging a prominent career
in the railroad and steel industries. In 1873 he became the managing partner of the
firm of Carnegie, McCandless & Co., and he had charge of the building and operation of the Edgar Thomson Steel Works until October 1, 1879 (ESWP 1892). He
also kept his position as Vice-President of the Allegheny Valley Railroad. The Edgar
Thompson works used the innovative Bessemer process, which Carnegie had
learned about in Europe. Shinn was credited with introducing many cost savings in
steel making, and the added profits made him near and dear to Andrew Carnegie.

In 1868 Shinn became a member of the ASCE, and in 1875 he joined the American Institute of Mining Engineers eventually becoming its President in 1880. In 1879 Shinn resigned as general superintendent of the Edgar Thompson Works with the intention to keep his stock shares, but Carnegie insisted he sell his major stake in the company (Nasaw 2006). I doubt that Shinn was overly perturbed at the large profit he thereby claimed. He was involved with many other businesses. From 1888 to 1891, which includes the year he was ASCE president, Shinn was Vice-President and General Manager of the New York and New England Railway Co. In the last two of those years he was President of the Norwich and New York Transportation Co. (ESWP 1892). Shinn was elected as one of five ASCE directors for 1889. One of his papers before the ASCE was considered by the railroad men the best ever written on railroad management and was awarded the Norman gold medal. So indeed, William Shinn was a top steel and railroad man from Pittsburgh.

I find it very interesting, and not mere coincidence, that Shinn was nominated to become ASCE president just 5 months after the Johnstown flood and while the ASCE investigation of that disaster was ongoing. At that time being nominated virtually assured election to the ASCE office. As he accepted the mantle of the Presidency from Max Becker the following January, Shinn could have insisted that the South Fork investigation be published. On the other hand, Shinn as President could have forbidden its release and let Becker bear the historic blame for sitting on it. I have little doubt that if Andrew Carnegie and Robert Pitcairn wanted the report kept sealed, Shinn would have gone along with that. And so would Becker, voluntarily or otherwise.

On May 5, 1892, less than a year after the South Fork report was finally released, William Shinn was dead at age 58 (ESWP 1892). Henry Frick immediately wired the news of their mutual friend to Carnegie in Sunningdale, England. All three, along with Robert Pitcairn, had also been members of the exclusive Duquesne Club of Pittsburgh. And Frick, Carnegie, and Pitcairn had been members of the South Fork Fishing & Hunting Club.

At the Cresson meeting James Francis spoke with a reporter and did not mince words. He was not happy with the delay in releasing the South Fork investigation report. From the June 30, 1890 edition of the *Johnstown Daily Tribune* (reproduced in entirety in Appendix 5), "Engineer Francis says the report will shortly be made public." I can only imagine how furious he would have been had he known the railroad men would suppress the report and keep it "secret" for another year! The article went on to say:

> In speaking of the matter yesterday [Francis] said he wished the report would be made public, as this thing of holding it so long was getting stale. He expressed as his opinion, however, that it would not be given out until the pending suits regarding the disaster were settled. Whether the reports would be favorable or not to the South Fork Club Mr. Francis did not say, but he did not hesitate to say that he did not approve of holding back the report so long. It was held, he says, at the instance of Mr. Becker, who first introduced the idea that they should not permit it to influence pending litigation, and the other members of the Committee had deferred to him. Mr. Francis is now in favor of having the report made public, but does not care to argue against the wishes of Mr. Becker, and as no other member of the Society cares to push the matter, it will be let alone. Mr. Becker is the engineer of the

Pittsburgh, Cincinnati & St. Louis Railway, and resides in Pittsburgh. The opinion prevails that it is on account of his business associates in that city that he takes his stand.

So there it was. Becker insisting the report be held back and the prevailing opinion, from Francis or from others at Cresson, that Becker "takes his stand" because of his Pittsburgh associates. His associates would have been railroad men, not the least of whom was Robert Pitcairn. It was surely no coincidence that Pitcairn and his assistant, Michael Trump, were on the organizing committee for the ASCE convention in Cresson. The meeting included a railway tour of the Conemaugh valley but I find no evidence of organized visits to the failed dam. If conference engineers wanted to go there they had to make their own way.

A sarcastic note about the engineers' "railway tour" from Cresson to Johnstown appeared in the *Johnstown Daily Tribune* (6.30.1890), in reply to a small note in a Pittsburgh paper:

The annihilation of Johnstown was proof enough to most people of the faulty construction of the South Fork dam, regardless of what engineers may report on the subject (*Pittsburgh Commercial Gazette*) - If the engineers have a free ride over the P.R.R. and are dined and wined free as the Pittsburgh jurymen and Judge were, and are of no better material, it is not hard to tell what their "opinion" will be – probably that the Johnstown people were responsible for the loss of fish in the dam under the control of the Pittsburgh Codfish, and that the heirs of those who were drowned should be made pay for the destruction of the little fishes [*Tribune*].

Several civil court cases against the South Fork Fishing & Hunting Club in Pittsburgh were not going well for the plaintiffs, and in Johnstown the folks were thinking "the fix was in." In one case Mrs. John A. Little sought compensation for the loss of her husband, the 43-year-old father of eight children she struggled to support. Her lawyer was met in court by the powerful attorneys Philander Chase Knox and his partner James Hay Reed. Both just happened to be Club members. I found no documentation about who served on the appointed jury. But I have no doubt that the jury included enough railroad and steel workers who enjoyed a living wage, so long as a non-guilty verdict (or no verdict) would follow. That case went on for years (McCullough 1968). No damages were ever awarded against the Club or its members.

Cresson, being close to South Fork and Johnstown, may have originally been chosen as the site for the convention with the intent of then releasing the investigation report and organizing visits to the dam for ASCE members. That might have been the intent, but still Becker and President Shinn kept the report under wraps. By coincidence Andrew Carnegie formerly had a favored summer home in Cresson.[2]

No trips to the South Fork dam were organized for this Cresson meeting, but some engineers, such as P. F. Brendlinger and Francis Collingwood,[3] took the

---

[2] Carnegie and his mother Margaret spent summers at their cottage in Cresson over a period of 12 years. Margaret died there on November 10, 1886 when Carnegie himself was still seriously ill from typhoid fever. His brother Thomas died of pneumonia the previous month. After losing his mother and brother in 1886, Carnegie said goodbye to his beloved home in Cresson (Nasaw 2006).

[3] Collingwood was a New York engineer who worked for a time under Col. Roebling and also served on the Croton Aqueduct Commission. He was Secretary of ASCE from 1891 to 1894. Died 1911 in New Jersey.

chance on their own to visit the breached dam, just 10 miles away. Brendlinger took his own measurements and presented detailed observations a year later at the Chattanooga convention.

Although many members at the Cresson meeting clamored for the South Fork report, clearly "pressing" for it, the investigation was not presented until the next annual convention, and not published in ASCE's Transactions until June 1891, 2 years after the flood. The reason given by Max Becker for sitting on the investigation report for 2 years was "…we do not want to become involved in any litigation…" But another suit was filed after the report was published, as declared in *Eng. News* (1891c), June 27, p. 607:

> The JOHNSTOWN FLOODS have not yet subsided, at least in the courts. Col. W. D. Moore has commenced suit against the South Fork Fishing and Hunting Club, that owned the dam, for the damages sustained by some of his clients. He charges criminal negligence and says the club membership represents $100,000,000 [net worth] and is able to pay.

Moore's client was Jacob Strayer of Johnstown, a dealer in lumber. Their intent was to sue not only the Club but also Col. Elias Unger. But for many reasons it never came to trial. The Club itself was bankrupt and insolvent by 1891, and various parcels were sold off or eventually went for sheriff's sales. Col. Unger had a heart attack in 1895 and died in Harrisburg less than a year later.

7.3   Annual Convention in Chattanooga, TN, May 1891

In 1891 ASCE's annual convention was held on Lookout Mountain, high above Chattanooga, TN.[4] By then ASCE had another President, Octave Chanute. He had long been a promoter of the ASCE and the engineering profession. He had known the outgoing President Shinn for many years, having collaborated with him on projects in the 1870's (Short 2011). For the first 5 months of Chanute's term the South Fork investigation remained sealed and in the hands of Max Becker. But that was about to change. *Eng. News* (1891a) had announced on May 2nd that, finally, a report would be given by the special committee on the Causes of the Failure of the South Fork Dam. For New York attendees, numbering about 42, a special train left Grand Central Station at 10 a.m. on May 19th. Cars were added and conventioneers picked up along the trip west. After arriving at Chattanooga they transferred to a local train that climbed Lookout Mountain under the stars in moonlight. The venue was the Lookout Inn, with the meeting commencing on May 21st. Chanute gave his Presidential address on May 22nd. Curiously, his very long speech made no mention of the Johnstown flood report that was to be released at that meeting.

The business part of the meeting began on May 23rd and included presentations by special committees. A brief summary of the South Fork presentation is given by

---

[4] Ironically, earlier that year, on February 17th, Johnstown had flooded again when both the Little Conemaugh and Stonycreek Rivers inundated the business part of the rebuilt city by several feet.

Eng. News (1891b) in their June 6th issue, p. 542. The summary says that "The committee on the Causes of Failure of the South Fork Dam presented its report...", but in fact only James Francis was there. The published list of members attending does not include Fteley, Worthen, or Becker. There is no evidence that they were present but somehow left off the list, as their names do not arise in any discussions of papers. All three men with Francis had attended the 1889 convention, but William Worthen did not come to Cresson in 1890. He may have had pressing business to attend to. But he might not have attended given advance word that the South Fork report would not be presented. He was undoubtedly as concerned about the delay as Francis was. Then in 1891 only Francis took the train to Chattanooga to present the long anticipated report. The absence of his co-investigators is evidence that Worthen and Fteley, close friends, stayed away as a "message" to ASCE's members that they were not pleased with how the report had been handled and kept secret. And Becker likely was absent to avoid unpleasant questions after the word had gotten out that he was the one sitting on the report so long.

How convenient for the likes of Robert Pitcairn and other Club members that the unsealing of the South Fork report did not happen in Cresson, but instead in Tennessee, far from Johnstown. The many eminent engineers who would hear the presentation could not easily afterward visit the dam to see it with their own eyes, to judge the conclusions based on their own knowledge and experience. And of course erosion and sightseers tramping over the remnants had altered the site in the span of 2 years so that all was not quite as it was when Francis, Worthen, Fteley, and Becker surveyed the breached dam.

With regard to history and the ASCE in 1889–91, it is most unfortunate that the stated concern about possible litigation took precedence over the higher duty to report to the profession and the public about safety issues that could then have existed at other major dams, both within and outside the U.S. Other towns and cities could have been at grave risk from dam failures. Then as now, public safety should have overruled all other considerations. But for Club members like Pitcairn, their reputations were more important than the safety of other dams. That was a motive for them to meddle in the release and content of the report on the South Fork dam.

7.4   Findings of the South Fork Dam Investigation

James B. Francis presented the basic findings at the annual convention of the Society in Chattanooga. The long-awaited report by him, Worthen, Becker, and Fteley was published in ASCE's *Transactions* in June, 1891 (v. XXIV, p 431–469). The last 9 pages included some of the report figures along with discussion material from the Chattanooga meeting. In the end the report stated that the dam would have failed even had it been maintained within the design as originally built in the 1850's. The committee concluded that:

"The [South Fork] Hunting and Fishing Club [sic], in repairing the breach of 1862, took out the five sluices [drainage pipes] in the dam, lowered the embankment about 2 feet, and subsequently, partially obstructed the wasteway [spillway] by gratings, etc., to prevent the escape of fish. These changes materially diminished the security of the dam, by exposing the embankment to overflow, and consequent destruction, by floods of less magnitude than could have been borne with safety if the original construction of 1851-1853 had been adhered to; but in our opinion they cannot be deemed to be the cause of the late disaster, as we find that the embankment would have been overflowed and the breach formed if the changes had not been made. It occurred a little earlier in the day on account of the changes, but we think the result would have been equally disastrous, and possibly even more so...." (Francis et al., 1891, p. 456).

This committee's claim that the dam, even as originally constructed, would have failed bears close scientific scrutiny. Two colleagues and I analyzed scientific aspects of the dam breach at South Fork (Coleman et al. 2016). We analyzed the time of concentration for the drainage basin and flood inflows to the lake on May 30–31. Eyewitness reports of the dam breach were reviewed in detail. We also examined whether two spillways (an original auxiliary spillway on the southwest abutment was missed by the committee) and the drainage pipes together, along with greater storage capacity behind a higher impoundment, would have prevented over-topping of the dam had it not been lowered as much as 0.9 m when the South Fork Fishing and Hunting Club rebuilt the dam. Our analysis was supported by river level observations that stream inflows to the lake had peaked hours before the dam breach. The calculations and results, presented in Chaps 10 and 11, reveal that the Johnstown Flood should never have happened in 1889.

There are a number of problems with the report by Francis et al. (1891). Foremost of these - it is inconceivable that hydraulic engineers of the stature of James Francis, William Worthen, and Alphonse Fteley would have overlooked the presence of a second spillway, on the western abutment of the dam. How did they miss that? These were eminent professionals who took survey measurements themselves of the dam remnants. The fact that the southwest abutment was lower than the dam crest was obvious from their own data. Perhaps President Becker exerted iron con-trol over the content of the report, instructing Francis, Worthen, and Fteley to ignore the abutment area in their calculations. If that happened they should have stood their ground. But they could hardly have appealed to higher authority. The president now was William Shinn.

I decided to look into the history of all four engineers who investigated the South Fork disaster; their careers, experiences, and family lives. Who were they and what kinds of "pressures" were they dealing with? And what do we know today about John G. Parke, Jr., the young engineer who witnessed the breach of the South Fork Dam? Little was known about his life before 1889 and he virtually disappeared after the Johnstown Flood. But the story of Parke's family and professional life can now be written.

References

Coleman NM, Kaktins U, Wojno S (2016) Dam-breach hydrology of the Johnstown flood of 1889 – challenging the findings of the 1891 investigation report. Heliyon 2, 54, https://doi.org/10.1016/j.heliyon.2016.e00120 [Accessed 16 Apr 2018]

Courier-Journal (1889) Newspaper edition June 4 1889, Louisville, KY
ENARJ (1889) Presidential address by Max Becker, Engineering News and American Railway Journal, June 29
Eng. News (1889a) Engineering news, June 8
Eng. News (1889b) Engineering news, June 15
Eng. News (1889c) Presidential address by Max Becker at the 1889 annual conv. in NJ Engineering News, June 29
Eng. News (1889d) Comments by a. Fteley, *Engineering News*, June 29 p 603
Eng. News (1889e) Engineering News, September 4
Eng. News (1889f) Engineering News, September 14 p 259
Eng. News (1890a) Engineering News, January 18 p 58
Eng. News (1890b) Engineering News, April 19 p 371
Eng. News (1890c) Engineering News, June 7
Eng. News (1890d) Engineering News, June 21, p 586
Eng. News (1890e) Engineering News, July 5 p 10–11
Eng. News (1890f) Engineering News, July 12 p 34, 35
Eng. News (1890g) Portrait of William Shinn, Engineering News, December 20
Eng. News (1891a) Engineering News, May 2
Eng. News (1891b) Engineering News, June 6 p 542
Eng. News (1891c) Engineering News, June 27 p 607
ESWP (1892) Proc. Engineers society of western pennsylvania, May 17
Francis JB, Worthen WE, Becker MJ, Fteley A (1891) Report of the committee on the cause of the failure of the south fork dam. *ASCE Trans* XXIV:431–469
McCullough D (1968) The Johnstown Flood. New York, Simon & Schuster, p 302
Nasaw D (2006) Andrew Carnegie. The Penguin Press, New York, p 878
Short S (2011) Locomotive to Aeromotive - octave chanute and the transportation revolution, Univ of Illinois Press, p 360
Tribune (1890a) Johnstown Daily Tribune, June 27
Tribune (1890b) Johnstown Daily Tribune, June 30 [article reproduced in Appendix 5]

Chapter 8
Biographical Sketches of John G. Parke, Jr. and the Engineers Behind the ASCE Investigation

In 2011, Emeritus Professor Uldis Kaktins traveled to Maine to visit my family in the house we rented by the sea in Machiasport. May is a rainy month on that coast, and on overcast days we collaborated on 1889 flood research, snacking on the blue mussels, periwinkles, and soft shell clams we gathered at low tides when the sea retreated hundreds of meters. An entrance to the Bay of Fundy lay only 35 km to the east. One warm, sunny day we drove up to Lubec where we crossed the bridge to Campobello Island, Canada. The Atlantic beaches there are drifts of cobbles, rounded granitic and metamorphic rocks.

At one point Uldis climbed up on a whaleback, an outcrop of bedrock shaped by glacial action, and to a lesser extent by beach erosion and wintry freeze and thaw. Grooves had been etched across the top of the whaleback, carved by the slow progression of a massive continental glacier over this landscape. The whaleback was far from the ocean then because sea level was lower by about 120 m, 25,000 years ago at the climax of the Wisconsin glacial stage. We were discussing and debating the natural history of the island and surveying the glacially deposited soils exposed in storm-eroded bluffs behind the beach. My professor had spent many summers bushwhacking the pine, spruce, birch, and balsam forests and cranberry swamps of Maine, mapping the glacial geology. Uldis stepped to the highest part of the whaleback, shaded his eyes as he looked seaward across the surf-pounded shore, and out of the blue said,

"We need to know more about those engineers."

"Which engineers?" I replied.

"That John Parke fellow at the dam, and especially Francis, Worthen, and the other engineers who did that South Fork investigation. What were they like, what outside pressures were on them at the time? Can we know that after all these years? We're focusing on the science but don't know the people at all. We need to try to see it through their eyes. How did their careers and experiences shape their judgements?"

"Especially Francis." I said. "It looks like he was in charge of that Committee. But was he really? James B. Francis that is; one of his sons was also called James."

© Springer International Publishing AG, part of Springer Nature 2019
N. M. Coleman, *Johnstown's Flood of 1889*,
https://doi.org/10.1007/978-3-319-95216-1_8

"I've been to their town you know, Lowell. Just northwest of Boston where I grew up. We should go there and walk those streets that Francis walked and along the canals. After all, he designed much of the infrastructure that powered those mills. Best way to get to know the man and the engineer."

He was right. Unfortunately we could not call up James B. Francis on the phone or even send a telegram. He had died 120 years earlier. David McCullough, who in 1968 published the exceptional book, "The Johnstown Flood," had the good fortune to interview some survivors and witnesses of the 1889 flood. His book remains the singular work to read to understand what happened that day, in the years leading up to it, and in the difficult days that followed. McCullough describes various engineering reports of the disaster and referenced the 1891 ASCE investigation, but does not mention that the report by Francis et al. (1891) essentially exonerated the South Fork Fishing and Hunting Club from blame in the 1889 flood.

McCullough's 1968 book on the flood is primarily a work of history. Over the years almost no science has been done to analyze that terrible dam failure. Uldis Kaktins and I dedicated ourselves to fill the yawning scientific gap about this flood, to understand in technical terms how and why the South Fork dam failed. We were joined on this journey by Stephanie Wojno, who like me was a former student of Uldis'. We sought to gain new insights about the dam breach and the history of the ASCE investigation. But as with many things, saying it fast makes it sound easy. Hard work lay ahead, and in the end, like science at its best, delving into the history of the 1800's would involve painstaking detective work.

I eventually made that trip to Lowell, Massachusetts, to walk the streets that both James B. Francis and William Worthen walked, to enter the home where Francis lived, the same home where the artist James Abbott Whistler spent the first 3 years of his life. But sadly this trip was not with my dear friend and professor. He had passed away a year earlier, cancer depriving us all of this gentle and good-humored genius. How much more could I have learned, touring that city of canals, a centerpiece of the Industrial Revolution, the "Venice of America," with Professor Emeritus Uldis Kaktins?

In the end it would prove important to delve into the history of all the men who served on the investigation, not just James Francis. It had seemed that Francis would have been the central figure in leading the committee, in drafting and finalizing their report and in handling its public release, which was delayed until 2 years after the flood. But Francis was not responsible for that delay or for all the perceived deficiencies in the report. What follows are biographical sketches for John Parke and all four of the ASCE engineers, with information about their careers, family histories, and aspects of their personalities. Interesting insights gradually emerge. Let us begin with the young engineer from Philadelphia who saw the South Fork dam overtop and witnessed its catastrophic breach.

8.1   John G. Parke, Jr., Engineer

John Grubb Parke, Jr. was born in Philadelphia on June 27, 1866. His parents were Thomas Hart Parke (1829–1876) and Anna Maria Torbert (1827–1897). The infant boy was named after his quite famous uncle, John Grubb Park. Some information is presented here about young Parke's uncle due to inconsistencies in historical refer- ences about their relation, as they had the same name and both attended the University of Pennsylvania, neither graduating there.

Parke's uncle attended Penn for one year, then entered the Military Academy at West Point where he graduated second in his class in 1849. He served in the Corps of Engineers where he performed extensive surveys in the western lands. Parke Sr. distinguished himself during the Civil War, rising to the rank of brevet Major General in the Union Army. He was present at numerous battles, including Ft. Macon, the Wilderness, Spotsylvania, South Mountain, Fredericksburg, Antietam, Ft. Stedman, Vicksburg, and the siege of Petersburg, where he assumed command of the IX Corp. General Burnside had been relieved of that duty after the poorly planned and executed Battle of the Crater. Parke Sr. joined the final pursuit of Lee's army until it surrendered at Appomattox. Near the end of his military career he was honored with the "plum" duty as superintendent of West Point from August 1887 to June 1889. On July 2, 1889 he retired after more than 40 years of army service. Parke Sr. died 11 years later at his home, No. 16 Lafayette Place, Washington, D.C. on Dec. 16, 1900. His remains were interred in Philadelphia (*NY Times* 1900).

Although it may seem odd today that an infant male would be named "junior" after his uncle, in the 1800's it was not unusual to name a male child after a renowned relative. Also, Parke Sr. and his wife had no children, so his brother Thomas Parke likely sought to honor his brother by naming his own son after the military hero.

That son, John Grubb Parke, Jr., attended the University of Pennsylvania, enter- ing as one of 69 freshmen in the 1882–1883 school year (Penn 1883). He skipped a year, being listed in the 1884–1885 Penn Catalogue of Announcements (Penn 1885) as one of 119 sophomores and was affiliated with the Towne Scientific School at Penn. The other "schools" at Penn at that time were the Department of Arts and the Wharton School of Finance and Economy. He entered class in the Civil Engineering Section as a sophomore. Parke was one of 101 juniors in the class of '87, with a home address listed as 403 North 33rd Street in Philadelphia at the inter- section of Baring Street, one block from today's Drexel Park. A remnant has been preserved from the original Parke house, still shown as #403, but additional struc- ture was added around 1930 so that today this is a multi-unit apartment building. It is a 15 min. walk from #403 to the Penn Campus. I have walked it - Parke had an easy amble to school.

He was a member of the Class Tennis Club and of the Alpha Tau Omega Fraternity, which then had 10 members at Penn. That fraternity was created by O. A. Glazebrook, E. M. Ross, and A. Marshall, at the Virginia Military Institute in 1865, founded on Christian rather than Greek principles, intending for its members to be free from partisan bias. Little else is known of Parke's school years at Penn, except that he played the role of a *Thracian*,[1] an inhabitant of the ancient region of Thrace, in the Greek play "The Acharnians."

Unrau (1980 p 87) incorrectly states that Parke Jr. was "…a recent engineering graduate of the University of Pennsylvania…" He left the school at the end of his junior year without earning a degree (personal communication: Archives, University of Pennsylvania, 2017). Parke worked with the Construction Department of the Pennsylvania Railroad from 1886 to 1888. Then in the spring of 1889 he was hired by the South Fork Fishing and Hunting Club to oversee plumbing work and construction of a sewer project on the Club property. The sewer work began in April in order to be finished in time for the summer season. But the failure of the South Fork dam at the end of May ended that work and the days of the South Fork Fishing & Hunting Club were over.

I have not determined how John Parke, Jr. landed his job at Lake Conemaugh. At that time a degree was not needed to become an engineer, especially for a young man like Parke who had studied civil engineering for 2 years at Penn and had very prominent family connections. The uncle after whom he was named had a long association with the railroads. Young Parke might have gotten his 1889 position at the Club through contacts from his first job with the railroad. Robert Pitcairn was a founding Club member and had been Superintendent of the Western (Pittsburgh) Division of the Pennsylvania Railroad for almost 25 years. During the Civil War one of his responsibilities was the arduous task of coordinating the transport of army personnel and materiel in western Pennsylvania. He met and got to know many army officers at that time, which may have included Parke's uncle who was now the West Point superintendent.

So where did John Parke, Jr. (Fig. 8.1) end up after his traumatic and unforgettable experience at the South Fork Fishing & Hunting Club? 2 months after the flood he wrote a letter to the ASCE committee, posted from Americus, Georgia. As of 1890 he was back from the South and living in Pittsburgh at 253 Shady Lane. I have not found where he was employed then. A university class book describes his later career path (Penn 1907). From 1897 to 1899 he worked with a civil engineering firm in Pittsburgh that was housed in the Westinghouse building. In 1899 he became Chief Engineer of the Union Steel Company at Donora, located across the river from Monessen, PA. That work involved construction of an open-hearth steel plant and blast furnaces and the building of rod, wire, and nail mills. He also oversaw railway work. On Thursday, August 8th, 1901 Parke married Miss Ella Etta Frantz, who hailed from the small town of Coal Center, PA, six miles south of Monessen. After Union Steel was bought by the U.S. Steel Company in 1905, Parke moved to Paterson, N.J. to become Engineer of Construction and then Chief Engineer of the Passaic Steel Company. In Paterson, Ella gave birth to their only child, Frantz Torbert, on April 17, 1906 (Penn 1907), one day before the devastating 1906 earthquake in California.

---

[1] The Iliad refers to *Thracians* as allies of the Trojans in their war against the Greeks.

Fig. 8.1 Images of John Grubb Parke, Jr. Left: From his youth at the University of Pennsylvania. Center: Parke in 1907. Right: Photo from his university class yearbook in 1922. Images reprinted with permission, University Archives, University of Pennsylvania (Philadelphia, PA)

In 1907 Parke was appointed Chief Engineer of the Pittsburgh Steel Company at its works in Monessen, which had begun operating in 1901. He held this position through the rest of his working career. He was a past Master of the Mason Lodge (#626) in Monessen and also a member of the Elks Lodge. Parke wrote several articles for engineering societies. He was a member of the Association of Iron & Steel Electrical Engineers, the Board of Trade in Monessen, the National Rivers and Harbors Congress, and the Americus Republican Club in Pittsburgh. The Americus Club was private but not financially exclusive; $6 covered annual dues for members in 1890. Several Americus Club members had also belonged to the South Fork Fishing & Hunting Club, including Lewis Irwin and life members Henry Frick, James M. Schoonmaker, James Hay Reed, and John Weakley Chalfant.

As of 1907 the John Parke family lived at 23 Knox Avenue in Monessen. By the time of the 1910 census, May 5th, he had been married to Ella (b. 1875) for 8 years and the family was renting a house in Monessen's Ward 1 at 104 First Street with their son and housekeeper Ellen Berkovitsch. At the 1920 census they lived in another rented home, 16 Schoonmaker Avenue in Monessen. Their son Frantz was then 14 years old. A nurse named Ellen Cypher resided with them but it is unclear whether she was simply a boarder or had been hired to provide care or assistance in the Parke home.

Parke's classmates from Penn had last seen him at their 35th reunion in June, 1922.[2] The reunion record noted that his son planned to attend Penn that very year, entering the engineering program like his father had. Parke penned some humorous thoughts for the reunion book (p 94 of Penn 1922):

Parke asserts that his chief mental diversion, physical pleasure and hobby are the three "R's" of reading, resting and riding; and he gloriously prances into the arena of honesty – we fear hypocritically sidestepped by some – by saying that his chiefest weakness is "look-

---

[2] At that time at Penn, records of male students were maintained in class yearbooks even if they did not graduate.

ing at the girls," and that "since short skirts came in, it was two years before I discovered that they bobbed their hair." Parke's message has a sort of a homespun and satisfying flavor to it: "Just working hard and enjoying life and trying to keep a sixteen-year old son from being a son-of-a-gun, and teaching him to be what his father was not."

In that last comment Parke may have hoped his son would study engineering and eventually earn a degree, as Parke himself had not. His young son never did start classes at Penn, but he sought an engineering career at the Carnegie Institute of Technology in Pittsburgh, residing in a dormitory there. Through a merger in 1967 that Institute eventually became Carnegie Mellon University. Sadly, early in 1925 Frantz Parke was struck by meningitis, entered Presbyterian Hospital in Allegheny City, and died 3 days later at noon on March 9th. He was interred at Howe Cemetery in Monessen. Five years later, word spread that Parke himself had become ill, early in 1930, with a condition that became life-threatening over the next 2 years. He was hospitalized for months but could not recover. His final days were spent in nursing care at his home, suffering from "cardiac insufficiency" and bladder cancer which had been diagnosed 3 years earlier. John Grubb Parke Jr. passed away at 162 Schoonmaker Ave. in Monessen, during the depths of the Great Depression. He was 66 years old on that Sunday, January 29, 1933.

The University of Pennsylvania printed an article on Nov. 1, 1933 about the passing of John Parke in its alumni magazine, "*The Pennsylvania Gazette*" (Penn Gazette 1933). The article incorrectly states that his work in 1889 dealt with "strengthening a dam near Johnstown, owned by a Fishing Club." In fact he was designing and constructing a sewer drainage project. On the day of the flood he was directed to work on the dam by a frantic Col. Unger in the face of rapidly rising water in Lake Conemaugh. Unger then sent him to South Fork to telegraph a warning about the dam.

The Penn article also disclosed that two of Parke's classmates from Penn, Harry Adams and Dan Stackhouse, were in Johnstown during the flood (Fig. 8.2). Stackhouse was from Johnstown, born in 1866 to Powell Stackhouse and Lucy Roberts. He narrowly escaped with his life but Adams was killed. The list of the dead does not include a "Harry" Adams, but one body found was Henry Clay Adams. At that time "Harry" was a common moniker for young men named Henry, a trend from the nickname for the many English monarchs named Henry. Henry Clay Adams was indeed lost in the flood. In Johnstown's records he is among nearly 200 identified victims with no record of burial. But that is true no longer. I will have more to say about Henry Adams in the chapter titled "End Notes". Henry outlived his father, Greenfield Adams, by only 5 years. Greenfield Adams had been a Circuit Court Judge in Kentucky and twice served as a Congressman from that State.

As a young man John Parke Jr. witnessed one of the worst disasters in U.S. history. Despite his inexperience he documented the events at the South Fork dam in remarkable detail. In his later career he played an important role in the golden age of iron and steel in western Pennsylvania. The engineering works he helped to build crafted the steel to make machines and tools that powered the United States and allies to victory in World War II. Parke's wife Ella survived him by 13 years and

Fig. 8.2 Daniel Morrell Stackhouse (*left*) and Henry Clay Adams were classmates of John G. Parke, Jr. at the University of Pennsylvania. Adams was class President the sophomore year. He had a degree in mechanical engineering and worked at Cambria Iron Company. Stackhouse also worked there, as Assistant to the Superintendent of Blast Furnaces. Adams perished in the flood. Images reprinted with permission, University Archives, University of Pennsylvania

witnessed the end of that war. She had been retired at 5625 Wilkins Avenue in Pittsburgh. Ella Frantz Parke suffered a stroke on October 2nd, 1946 and passed away just after midnight the following day.

Looking back to his Pittsburgh days, John Parke is mentioned in an article about the Bellefield Arch Bridge, a beautiful stone arch 150 ft. across at the entry to Schenley Park. That bridge connected the Carnegie Library Building to baseball's Forbes Field. Parke was one of the inspectors of the bridge's masonry work. He was credited with developing an efficient graduated square for placing the voussoirs, which are the wedge-shaped stones that form the key support structures in an arched bridge. The stonemasons on that job gave his setting square a rather affectionate nickname, "Parke's Fiddle." Sketches of the Bellefield Arch Bridge and "Parke's Fiddle" appear in an issue of *Eng. News* (June 22 1899).

Today the Bellefield Arch Bridge remains in place, but you'll never find it. The gorge under the arch, once known as St. Pierre's ravine, was used to dispose of a mountain of excavation and construction debris. By 1915 the gorge was filled and the bridge itself covered over. This was truly a sad fate for such a magnificent bridge, but proof of the rapid growth of downtown Pittsburgh. Schenley Plaza and its beautiful bronze and granite fountain, named after Mary Schenley, were built atop the site at the entrance to Schenley Park. In 1965 the Frick Fine Arts Building, part of the University of Pittsburgh, was built directly behind the fountain and over the filled ravine and buried bridge. That building was named after Henry Clay Frick, a prominent and charter member of the South Fork Fishing and Hunting Club.

8.2   James B. Francis, Jr., Lowell's Hydraulic Engineer

The life of the internationally renowned engineer, James Bicheno Francis (Fig. 8.3), was chronicled by his friends and colleagues (ASCE 1901; Fitz Gerald et al. 1894; Frizell 1894; Mills 1892; Worthen 1893) and in letters to and from his son James. He was born on May 18, 1815 in South Leigh, which today is a tiny village west of Oxford, in West Oxfordshire, England. In 2011 fewer than 240 people lived in South Leigh parish. The village lies near Limb Brook, a small tributary that flows eastward through gently rolling countryside to the River Thames. The hamlet is surrounded by farmland, hedgerows, and stone walls.

James Francis' parents were John Francis and Elizabeth Frith. Elizabeth was born on May 7, 1790 in Greenham, England, a village west of London. She was the daughter of James Bicheno and his wife, Ann. Interestingly for that era, James Bicheno made special arrangements in his last will and testament that gave Elizabeth independent control of her inheritance separate from her "...present or any future husband." Even so, she and her husband were of modest means and their son's education was limited. Young Francis had early schooling until he was 14 years old. He never went to a university but more than made up for this with an inquiring mind and exposure to various and innovative engineering projects at an early age. The young lad was aware that his lack of formal education, especially in mathematics, would be a disadvantage in the engineering profession. He overcame this by

Fig. 8.3 Image and signature of James B. Francis (Source: *Eng. News*, Jan. 1, 1887d)

concentrated self-study. In common with the best scientists and engineers, then and now, Francis stayed at the cusp of his profession by always learning and experimenting. He became a "student" for life.

In 1828 Francis' father became superintendent of a short rail line in southern Wales, the Duffryn, Llynwi & Porth Cawl Railway. The elder Francis also supervised construction of the Porth Cawl Harbor Works in Glamorganshire, South Wales. Young James must have been little challenged at school because in 1829 he applied for an apprenticeship in the work. At age 14 he was engaged in the harbor works associated with the railway, under chief engineer Alexander Nimmo. This experience in the new and developing railroad technology was a boon for this bright young man, akin to being part of the space program in the U.S. in the 1960's. In 1831 he was working on the Grand Western Canal in Somersetshire and Devonshire under chief engineer James Green. Young Francis could not have known then that most of his working career would be devoted to canals, mills, and water power (Worthen 1893; *Eng. News* (1887c).

In a letter to John Comstock of Dartmouth College (11/29/1892), Francis' son James reviewed details of his eminent father's life. His father emigrated to the U.S. in 1833, arriving in New York City on April 11th with letters of introduction to Phelps, Dodge, and Co. He sought out the well-known engineer, George W. Whistler, whose future son would become the eminent artist, James Abbott McNeill Whistler (b. July 11, 1834). Francis was hired to work in Stonington, Connecticut as an assistant to James P. Kirkwood on the New York, Providence, and Boston Railroad (*Eng. News* (1887d). In 1834 Whistler began building locomotives in New England.

Both Whistler and Francis moved to Lowell, MA in 1834, where Whistler became Chief Engineer of the Proprietor of the Locks and Canals Company on the Merrimack River. Francis found work as a draftsman in that company's machine shop and was soon recognized for the care and precision of his work. One of Francis' first tasks was to dismantle and measure the parts of a locomotive that had been shipped from England. His detailed drawings of those parts produced plans for engines to be used on the earliest New England railway, the Boston and Lowell (*Eng. News* (1887d). Francis was soon working on mill designs, and at age 19 was given responsibility that included construction oversight for the Boott Cotton Mills in Lowell (Fitz Gerald et al. 1894). He afterward got involved in the dam, canal, and hydraulic work that supplied power to the many mills that were built in Lowell. This interesting work and numerous consulting projects involving dams, turbines, and canals occupied him for the rest of his life.

In 1837 George Whistler left the U. S. to work on railroads in Russia. He sold his home at 243 Worthen Street to James B. Francis and before leaving appointed him as Chief Engineer of the Locks & Canals. In July that year Francis married Sarah Wilbur Brownell (1817–1904), daughter of George Brownell of Lowell, originally from Waltham, MA. Brownell was Superintendent of the Locks and Canals Machine Shop where Francis had been working. Brownell had come to Lowell from the cotton mills previously built at Waltham. He was a top expert in factory machinery. By this time Francis knew that he needed a greater knowledge of mathematics to support the hydraulic engineering work. His intense self-study of mathematics gave

Francis knowledge of math that, according to Worthen (1893), was beyond the college requirements of that day.

In 1845 the "Locks and Canals" company sold their machine fabrication branch to a new corporation, the Lowell Machine Shop. At that time the main manufacturing corporations became proportionate stockholders of the Locks and Canals Company. For the next 40 years James Francis worked on, improved, and cared for the Pawtucket Dam, the canals, and the distribution of water power. He also served as the consulting engineer to factories which grew to be among the largest in the world (Fitz Gerald et al. 1894).

In 1846 Francis was appointed Agent of the Locks & Canals Company. Three years later the Manufacturing Companies of Lowell sent him on a research trip to England to learn about timber preservation methods. In 1852 Francis was elected[3] a member of the American Society of Civil Engineers (ASCE). Even though he was not one of the 12 engineers who first met to form the Society, he was a founding member because he was elected in its first year of existence.

The Northern Canal was built in 1847 to satisfy the growing power needs of the mills. Its design and construction was overseen by James B. Francis and, when completed, it provided a more equable means to distribute water power among the companies in Lowell (Worthen 1893). The flow into this canal is controlled from the Pawtucket Gatehouse at the dam on the Merrimack River. Ten gates are used to regulate flow into the Northern Canal, which has branches that lead to the other canals. These gates were lifted and lowered by long screw rods powered via belts and gears by a Francis turbine. A lock chamber gave access to the Northern Canal on the southeast end of the gatehouse. As of 2017 the National Park Service (NPS) was giving tours of the dam and the gatehouse.

Engineers like Charles Whistler, Charles Storrow, Uriah Boyden, and young companions like William Worthen inspired Francis to make his own advances in the theory of hydraulic engineering. His best-known publication was "Lowell Hydraulic Experiments," which came out in four editions: the first in 1855, and a considerably expanded second addition in 1868. The 3rd and 4th editions were issued in 1871 and 1883. Francis became internationally recognized for this treatise on hydraulics. The underlying research is beautifully documented in reference materials, including work diaries and logbooks maintained in Lowell by the NPS. An 1865 paper, also well known, was titled "The Strength of Cast Iron Pillars."

In the 2nd edition of his hydraulics treatise, Francis describes how water power was distributed at that time. Grants had been given to eleven manufacturers in quantities he calls "Mill Powers." Each mill power was a little less than 100 horsepower. He gave an example of the grant to the Merrimack Company (Francis 1868, p. x),

---

[3] In 1889 ASCE changed how new members were admitted. Before that year applicants were subject to being "blackballed" by existing members. The change in 1889 is discussed in Appendix 6. In some professional societies today, a relic of preferred election survives in the selection of "fellows."

which reveals not only the length of an operating day at that mill but also the care he used to equitably share power:

> Thus, to the Merrimack Manufacturing Company, there have been granted 24 2/3 mill powers, each of which consists of the right to draw, for 15 hours per day, 25 cubic feet of water per second on the entire fall. Up to this time, there have been granted at Lowell 139 11/30 mill powers, or a total quantity of water equal to 3595.933 cubic feet per second.

Francis became a Vice President of ASCE in 1869. When word spread of the breach of the Mill River dam in 1874, the Society sent a committee of three civil engineers to Williamsburg, Massachusetts, to unearth the cause of the disaster. The three were James B. Francis, Theodore Ellis, and William Worthen. The investigation proceeded quickly and the report was published within a month of the disaster (Francis et al. 1874). They concluded that no engineer was responsible for the design and that it was the "work of non-professional persons." "The remains of the dam indicate defects of workmanship of the grossest character." I shall have more to say about this 1874 report by Francis et al. when comparing its conclusions to ASCE's review of the Johnstown flood.

Francis and his wife traveled to Europe in 1879 on an extensive trip intended for both business and pleasure. A travel document of the U.S. Department of State, dated Jan. 8, 1879, describes him then at the age of 63. He was 5 feet, 9 inches tall, had grey eyes, white hair, a high forehead, and an "ordinary" nose. Throughout their journey the Francis' wrote letters home of their adventures. The NPS has an archive about the history of Lowell, with bound volumes of letters related mostly to the younger James Francis. The archive includes correspondence with his father and a series of letters from places around Europe. I learned much about the elder Francis from those letters, his particular and thoughtful insights and way of thinking, and the things and experiences that impressed him as well as those that did not. He addressed his son James in these letters as "Colonel" or "Col," the title referring to his highest rank attained in the Union army. It is unclear whether the father wrote to all four of his children or corresponded mostly with son James, counting on him to pass on news of their travels.

Francis wrote a letter dated Jan 24th to his son James while aboard the steamship "Frisia." The ship expected to arrive shortly at Plymouth and then the Francis' would travel to Cherbourg. They spent considerable time in Paris, leaving there on February 2 and arriving the next day at the Hotel L'Europe in Lyons. Francis found this city "delightful." Their whirlwind trip continued through France to Italy, reaching Rome on Feb. 12, his wife's birthday. They were not happy with the room at their hotel and the next day found nice lodgings with a sunny view. Francis thought Rome was mostly "ruins and churches." They posted a letter on Feb. 25th from Rome, noting how hot the weather had been. A letter sent from Naples on March 12th took 19 days to reach Lowell. They visited many cities in Italy then went on to Switzerland, Germany, the Netherlands, England, and Scotland. A letter from Germany noted their trip to Cologne on a Rhine steamer, arriving on May 1st. A letter to son James on July 16th included a sketch of the profile of the dam at Edinburgh's Edgelaw Reservoir. They went west to Glasgow then took a boat to

Dublin and toured Ireland. They arrived in Bath on August 8th and returned to London on the 9th, where many letters from home awaited them. After visiting Liverpool they reached Ambleside on the 20th and had difficulty getting a room because a "great wrestling match" was to take place the next day. Their breathtaking trip would soon end with a return to the U.S.

Coming from the canal city of Lowell, the Francis' particularly enjoyed the waterways of Venice where boat travel efficiently took the place of wheeled contrivances. Francis noted how the gondoliers had a special call to avoid boat-to-boat collisions at sharp blind turns in the canals. Francis' letters have an interesting character – I found virtually no references to the food and drink available on their travels in Europe, and nothing related to art work, but there were comments on the architecture and transportation systems and the theater. A common theme was to talk about schedules of transportation, when they departed from and arrived at various places. In one letter Francis gave a litany of train schedules, as if that would be of profound interest to loved ones at home.

The following year, in 1880, Francis was elected President of the ASCE, the Society recognizing his accomplishments with one of its highest honors. A major part of his work in Lowell had been to maintain the dam, canals, locks, and turbines. He equably distributed the available water power at Lowell among the various manufacturers, mainly woolen mills and machine shops, according to their respective rights. Francis' career benefitted from an early association with the geniuses who together made Lowell and New England into industrial power centers. He knew Nathan Appleton, Patrick Jackson, the Lawrence and Lowell families, Kirk Boott, and others who laid the foundations for industrial growth.

8.2.1   The Francis Family

Seven children were born to James and Sarah Francis in their home on Worthen Street, but one male infant survived for only a day in May, 1844. The U.S. census of 1860 shows that eight family members lived in the Francis home, including James B. and his wife Sarah, sons George, James, Charles, and Joseph, and daughters Elizabeth and Lydia. Ten years later there were seven people in the household: James senior and his wife, 30-year-old son James, now a civil engineer, 17-year-old Lydia, who was an invalid, and three domestic servants named Mary Ryan, Mary O'Hare, and Alice Mehan.

James and Sarah's youngest son, Joseph, perished from cancer in 1869 at the age of 18 yrs., 10 mo. Their youngest daughter Lydia died of peritonitis in April, 1878. Four children survived after their father's death. His oldest son, George E. Francis, M.D., was born in 1838 (d. 1912). He had been in the U.S. Navy and became a well-known physician in Worcester, Massachusetts, married to Rebecca Kinnicutt. Daughter Elizabeth married Henry Bennett of Gloucester and was living in Bay View, Massachusetts.

Son James, born March 30, 1840, was a Civil War veteran who began the war as a Captain in company A, 2nd regiment, Massachusetts volunteer infantry, and rose to the rank of Lieutenant Colonel. He had genuine concern for the men who served with him. After the war he wrote sincere and persuasive letters to the Pension Commission on behalf of wounded veterans from his command. The younger James eventually succeeded his father as Agent and Engineer of the Proprietors for Locks and Canals. However, early in 1894, just 17 months after his father's death, an attempt was made to place James in a subordinate role to the consulting engineer for the management of water power. James' father had been the consulting engineer until his death. James refused to accept that role and in a private letter to Mr. Lyman, dated March 1st, offered to withdraw from his services to the Locks and Canals Company. He continued to be an accomplished hydraulic engineer in his own right. He passed away in the winter of 1898 from injuries sustained after falling from a carriage.

Son Charles was a Harvard graduate and, like his brother James, was also a war veteran. He had extensive experience with dams, hydraulic mining in California, and worked with the Mexican Central Railroad. He was a hydraulic and civil engineer in Davenport, Iowa, serving as Engineer of Public Works at the time of his father's death (Fitz Gerald et al. 1894). Like their father, both Charles and James were members of the ASCE, James having been elected in 1893 after his father's passing.

8.2.2   Francis' Retirement

On Jan. 1, 1885, James B. Francis resigned the office of Special Agent of the Proprietor of the Locks and Canals on the Merrimack River. He had held that position nearly 40 years, since Sep. 27, 1845. He was then appointed as Consulting Engineer and formally held that position 7 years until the time of his death. He had also served as director of the Gas Light Company in Lowell for 43 years and was involved with many professional, civic, and charitable organizations.

During his semi-retirement Francis served on the Lowell city council for 5 years and was elected to the Massachusetts State Legislature where he served for a year. Those political forays must have been exasperating experiences for an engineer like Francis, whose life was immersed in the logic of math and science. He continued to be an active member in the ASCE, attending and participating in many meetings and continuing his active role as Consulting Engineer and various civic duties. He had more time now for travel and personal pursuits. His interests included the history of Lowell and local flood records. In 1886 he wrote about great floods on the Merrimack River, which included a story of the greatest flood known up until October 1875. The dam at Pawtucket Falls had not yet been built, but based on an 1836 report by Uriah Boyden the height of floodwater above the crest of the 1886 dam reached ~13 ft., 5 inches. Significant uncertainty was expressed about this estimate. Another large flood occurred on April 23, 1852, which probably included significant snowmelt. This flood had a "…high-water mark at the Guard Gates of Pawtucket Canal [that] was fourteen feet one inch above the level of the top of the dam at the head of Pawtucket Falls." (Francis 1886).

One of Francis' consulting jobs was in Boston, advising on a project to improve drainage to prevent flooding in Roxbury. A report by Francis and other engineers was used to seek funding for the project from the legislature (*Eng. News*, Feb. 5, 1887, p 98). On May 5, 1887, before leaving with his wife for Europe on June 20th, Francis (1887) submitted a narrative about Kirk Boott and his experience in the British Army in Europe. The work was published by Lowell's Historical Association in July, 1887. Francis prepared the narrative with material he found in the library of the Boston Athenæum. Like Francis, Boott's father had emigrated to the U.S. from England. Boott played a tremendous role in the founding and operation of industries in Lowell. The NPS today has offices in the Boott Mill complex, where visitors can tour a museum that includes a restored weaving room with operating looms. Boott died in his carriage on Merrimack Street in Lowell on April 11, 1837, four years to the day after Francis first arrived in the U.S.

Francis and his wife left for another tour of Europe in the summer of 1887. He was at the pinnacle of his professional career and had obtained financial security; he now had the means to reward himself and his wife with a dream vacation. Their departure was announced in the "Personal" section of the June 25th, 1887 issue of *Eng. News*: "Mr. James B. Francis, Past President Am. Soc. C. E. sails from Boston for Liverpool, on the 20th inst., to be abroad some months." Francis stayed in touch because his son James had assumed his father's duties in the canal system and relayed information about the elder Francis' consulting work. For example, while they were abroad their son got a letter from a Wisconsin attorney seeking a deposition from the elder Francis. The son replied that his father was away for several months and it was unlikely he would be prepared to make the deposition.

Around this time the younger James Francis was having difficulty with his son Joe, referred to as "Little Joe" in his grandfather's letters from Europe. At that time son James and family lived on Andover Street, an east-west thoroughfare in Lowell.[4] "Little Joe" was 12 years old in 1887 and must have been a handful for his parents and grandparents. His father penned a letter to a Mr. Hale, asking him to take Joe into his family and arrange for his education and training. In the letter he says "Mrs. Francis calls Joe a nice boy, but nevertheless he is badly in need of proper discipline."

For a week starting on June 1, 1889 the newspapers were filled with stories about Johnstown. Later that week, James B. Francis and the other committee members traveled to the Conemaugh Valley to examine and survey the remains of the South Fork dam. The following month was a tragic one for the Francis family. Caroline, the wife of their son James, passed away on July 30, 1889. She was only 51 (b. June 19, 1838), the daughter of Franklin Forbes.

---

[4]Andover becomes Church Street after crossing the bridge to the west side of the Concord River.

8.2.3   Illness and Death of James B. Francis

The ASCE report on the South Fork dam was finished in January 1890 but was then sealed, not published until June, 1891. According to a family record, Francis was experiencing symptoms of an illness. One of the last projects he consulted on outside of Lowell was at Providence, Rhode Island, in June of 1890. Francis had been the original consulting engineer on the Fruit Hill Reservoir. When it began leaking he was called to suggest repairs to the City Engineer, a gentleman named Shedd. He traveled to ASCE's convention in Chattanooga in 1891 to present the South Fork report. The other investigators did not attend. Francis' condition became quite grave in 1892. He was unable to attend some meetings and word may have spread of his health troubles. The ASCE made him an Honorary Member in April. In a letter dated August 18, 1892, Francis's son replied to a Mr. Dexter of Boston, who had asked the assistance of the senior James Francis. The younger James replied:

> My dear sir:
>   If you feel that I can be of service, I will go to Biddeford to be there on the 27th, provided nothing occurs here to prevent. My father is the one who has been in the habit of advising, but now he is seriously ill and may never be strong enough to resume work of any magnitude.
>
>   Very truly yours,
>   *James Francis*

James Bicheno Francis died 1 month later, on September 18th, 1892 at the age of 77, just 15 months after the public release of ASCE's report on the Johnstown flood. His son James wrote that his father's cause of death "…was a tumorous growth in the bladder, which began probably two years previous to his death. A surgical operation was performed September 1st, from which he was unable to rally." The elder Francis had entered a private hospital in Boston and undergone what was considered radical surgery at the time for removal of the growth, which had been diagnosed with a newly invented primitive endoscope. The surgeon employed a metal tube as a prosthetic device, but in the end the outcome proved fatal for the brilliant engineer. His body was interred in the Lowell Cemetery.

James handled the affairs of his father's estate. In letters written at that time to his brother Charles and sister "Bessie" he signed rather formally using his full name, "James Francis."

8.2.4   The Legacy of James Francis

James B. Francis and my late professor and friend, Uldis Kaktins, evidently shared similar philosophies about our place in the universe. In Uldis' classes many students developed their first understanding about the true age of the Earth, and the sobering meaning of deep geologic time. A former employee of James Francis recalled a conversation between the renowned engineer and a minister, who had declared that "the future employment of men would consist in singing the praises of God." Francis replied to the cleric that God "must be a very stupid old fellow to care for so much praising." He asserted that religious doctrines, ordinances, and habits were chiefly "…valuable as giving employment to the minds of men and keeping them out of mischief." (Frizell 1894).

Francis' son James described several major engineering accomplishments of his father, including overseeing a greater part of the reconstruction of the Pawtucket Dam in 1875 and 1876, and the building of hydraulic lifts for the Guard Gates of the Pawtucket Canal in 1870. The emergency flood gate that was built here atop the Guard Locks is known as the "Francis Gate" and was intended to block the canal to prevent major flooding in Lowell's industrial district. It was initially ridiculed as a waste of money, called "Francis' Folly," until a great flood occurred in 1852, just 2 years after the gate was built. It saved the downtown from flooding, and did so again, twice in the 1930's and more recently in 2007 and 2008. I have drifted beneath the Francis Gate in a canal boat while traversing the lock on the Pawtucket Canal. The sliding gate itself has been replaced since the time of Francis. One of the years the Francis gate was dropped was 1936. That year the same regional storm inundated Johnstown, PA, and produced a flood level of 17 ft. at City Hall.

In a letter dated Nov. 26, 1892, sent to H. F. Miller of Lawrence, Massachusetts, James Francis listed in the following order the honors and memberships his father, James Bicheno Francis, had attained (source: Archives, Lowell National Historical Park):

- Member, Boston Society of Civil Engineers, July 3, 1848
- President, Boston Society of Civil Engineers, Apr. 27, 1874 - Aug. 7, 1874
- Honorary Member, Boston Society of Civil Engineers, May 20, 1891
- Member, American Society of Civil Engineers, Nov. 5, 1852
- Vice President, American Society of Civil Engineers, Nov. 3, 1869 to Nov. 2, 1870
- Vice President, American Society of Civil Engineers, Nov. 6, 1878 to Nov. 3, 1880
- President, American Society of Civil Engineers, Nov. 3, 1880 to Jan 13, 1881
- Honorary Member, American Society of Civil Engineers, April 5, 1892
- Fellow of the American Academy of Arts and Sciences, Boston, Nov. 13, 1844
- Honorary Degree of Master of Arts, Harvard College, 1858
- Honorary Degree of Master of Arts, Dartmouth College, 1851
- Member of American Philosophical Society of Philadelphia, April 21, 1865
- Member of Boston Society of Natural History, Jan. 1864 to Oct. 9, 1868
- Honorary Member, Arkwright Club, Boston, Jan. 1883
- Honorary Member, Winchester Historical and Genealogical Society, 1885
- Honorary Member of the American Society of Irrigation Engineers, Salt Lake City, UT, Sept. 1891

Over the years Francis contributed professional papers that added to the engineering knowledge base. These included "Experiments on the Deflection of Continuous Beams," "Experiments on the Flow of Water over Submerged Weirs," and "Experiments on the Preservation of Timber." Francis also wrote "Experiments on the Percolation of Water through Portland Cement Mortar," which was part of his paper titled "High Walls or Dams to Resist the Pressure of Water." Other articles examined the durability of iron water mains; the Provincetown dike; a translation of Darcy's experiments on the flow of water through cast-iron pipes; the concept that the maximum velocity of water flowing in open channels occurred below the surface; the distribution of rainfall over New England during the storm of early October, 1869; experiments on the Humphrey water turbine; and the effect of a rapidly increasing flow of water into a stream on the flow conditions below that point (Mills 1892).

Francis is particularly remembered today for inventing the highly efficient Francis turbine, a mixed-flow reaction turbine. Francis improved on earlier designs by Howd and Boyden. Details of the historic evolution of turbine design are described by Lewis et al. (2014). Turbines based on the Francis design are still operating at many dams today. For example, Francis turbines are operating at Hoover Dam in Nevada and at the Three Gorges Dam in China. The Francis equation for estimating the flow of water through rectangular weirs remains useful in the hydraulic engineer's toolbox. The many hydraulic experiments described in Francis' 1855 treatise provide evidence of a brilliant engineer's systematic approach to solving difficult problems. Francis also devised a sprinkler system for fire protection at the mills in Lowell. His original design would flood a large part of a building, including areas not touched by fire,[5] but this was a small risk compared to the extreme fire danger present in the mills.

8.2.5   James B. Francis as Remembered by Colleagues

To gain additional professional and personal insights about Francis, I reviewed memoirs about his life and career written by those who knew him best (Worthen 1893; Fitz Gerald et al. 1894; and Frizell 1894). Hiram Mills, who had known Francis for more than 36 years and worked as his assistant in 1856, also composed a memoir (1892). Was there anything about his personality that could have influenced his work on the committee to investigate the Johnstown flood? William Worthen had worked with him in Lowell starting in 1840, and later served with Francis on the ASCE investigations of the dam failures at Mill River and South Fork. Both men became Presidents of the ASCE. Worthen was a close friend of the Francis family, as he was addressed "affectionately yours" in a letter dated August 14, 1876 from Francis' son James. The younger James was writing back to Worthen because his parents were away, vacationing at breezy Nantucket during the height of summer. At the letter's close, James informed Worthen that "Lily is pretty comfortable and sends much love." "Lily" was a nickname for sister Lydia, who was very ill at the time.

---

[5] Not yet invented in the time of Francis, modern sprinkler systems use flow heads activated by heat that release spray only in areas affected by fire.

Worthen (1893) penned a memorial article about the life and career of Francis, published in the *Proceedings of the National Academy of Arts and Sciences*. He wrote that Francis showed himself "…in his published works and reports a close and careful investigator, suggestive in his methods, and of good judgment. Often chosen from his established integrity referee and commissioner, not only in the line of his profession, but outside, his decisions were without bias. As leading hydraulic expert of this country and often retained in suits, he never considered himself the attorney of his client, but gave his evidence honorably, agreeably to the facts and scientific precedents. In this his example is worthy of imitation…" "In private life Mr. Francis was honest, sympathetic, neighborly, not hasty in forming or giving opinions, but always consistent and decided…".

Eng. News also printed a reminiscence of James B. Francis, written by a man who had done canal and mill work for Francis for more than 10 years in Lowell, from 1857–1867. More than any other memoir, and beyond what I found in the family's letters, Joseph P. Frizell (1894) describes personal aspects of the famous engineer. He wrote that "Massive solidity was the characteristic of all Mr. Francis' designs. Safety the first consideration. Novelty or brilliancy he never aimed at." Francis was a very conservative man who gave little encouragement to inventors. He could see the merits of a new thing when they were unmistakable. When asked about the adoption of new things, Francis commonly replied "I am one of those stupid old fellows who can't be turned from the beaten path. You will waste your time in talking with me."

Francis had a quaint, subtle, dry humor that filled his conversation. He was quickly angered but had self-control of his temper. Frizell noted that "Often when things went wrong or when gross carelessness or remissness was apparent, the flashing eye, the clenched fist, the half-smothered oath gave notice of a temper not to be trifled with. But he instantly got a firm grip on it and proceeded to consider dispassionately what was proper to be done." On one occasion Francis felt that a company manager was dealing unfairly with him. He went to the manager's office to confront him, was ordered out, and promptly left. The manager decided to follow with a threat to do violence, but wisely retreated when Francis turned and faced him with a fierce look "foreboding no child's play."

Frizell had suffered a long siege of typhoid fever and recalled that Francis' "attention and kindness to me during this emergency made an impression on my mind that will never be effaced." One winter day in 1859 he witnessed an accident in which Francis slipped badly on ice, slid down a flight of stone steps, and struck the large arm of a water wheel. Had it been turning Francis would have been instantly crushed to death, and the history of Lowell and the study of hydraulic theory would have forever changed.

Francis not only worked to control his temper, but Frizell said he had the "happy faculty of restraining the expression of opinions which did not run in the popular groove and might involve him in discussion and argument, which he always aimed to avoid." This latter point is significant. If Francis sought to avoid arguments, this may have affected how he participated in and wrote about investigations, such as that at South Fork. We also know from a letter by son James that his father was experiencing health problems before the Johnstown flood report was finally published. The family was traumatized by the death of James wife on July 30th, the month after his father was appointed to the South Fork investigation committee. Francis visited the remains of the South Fork dam with the other committee members, but given what was going on in his life, questions can be raised about how much of his input went into the final 1891 report. There are appendices of weir and pipe flow calculations that he probably did, but Fteley or Worthen could as easily have done that work. The report does not give separate attribution to the contents. But in the end he alone presented the South Fork report to his Society at a meeting in Chattanooga.

8.2.6   Visiting Lowell Today

The story of the remarkable James B. Francis is best understood by visiting the town of Lowell, northwest of Boston. Lowell was a planned industrial complex, born in the region south of the Merrimack River and north of what is now the Pawtucket Canal. It was realized that the energy from the fall of water levels over a short lateral distance could be captured to run machinery on a large scale. The eventual construction of a dam at the head of Pawtucket Falls increased the water levels entering the canal and therefore sustained and ramped up the available water power.

Lowell is a beautiful historic town with its canals, museums, and statuary parks. Many of the works of James B. Francis have been preserved in the dam, the canals, the locks and existing mill structures. These are now part of the Lowell National Historical Park. Although it has been claimed that the industrial revolution began here, that is not strictly true. Nevertheless, Lowell became a powerful manufacturing center where upwards of 10,000 people worked in textile mills driven by hydraulic power. Lowell showed the way to the future, but slowly declined as electrification proceeded in the late 1800's. Electrification meant that industries would no longer have to be sited on rivers and would not be limited by the magnitude of available water power.

The NPS provides trolley tours and boat rides on portions of the canals. A tour on the Pawtucket Canal passes through the so-called "Guard Locks" and under the renowned "Francis Gate." I have visited the campuses of the University of Massachusetts at Lowell. The Francis College of Engineering is named in honor of James B. Francis and is a fine and ongoing tribute to his life's work.

Enough has been preserved of downtown Lowell to give the visitor a feel of what it was like to live here in the mid- to late-1800's. That was especially true early in the morning in August of 2017, when I found myself walking over a bridge that crossed the Concord River. At that hour Lowell was a quiet, sleepy college town with few cars on the streets. Near this bridge are the lower locks of the Pawtucket Canal. Many of the old mill buildings still stand, and the NPS maintains a visitor center and museum. Some of the mill buildings have been converted to modern businesses, apartments, and condominiums. On a sad note, the American Textile

Fig. 8.4 Left: Photo from NPS tour boat, approaching the downstream end of the Guard Locks, Pawtucket Canal. Right: View from upstream side of the Francis Gate. Gate is suspended over the center of the Guard Locks, located about 500 m downstream from the entry to the Pawtucket Canal. Just 75 m south of here the canal passes under Broadway Street. Note vertical notch in stone wall down which the gate slides when lowered. Photos in 2017 by the author

History Museum, formerly at 491 Dutton Street in Lowell, closed in 2016. Its board of trustees assumes the responsibility to protect the museum collections and to transfer them to appropriate nonprofit groups for their preservation and the public good.

The renowned "Francis Gate" at the Guard Locks on the Pawtucket Canal (Fig. 8.4) is credited with protecting the downtown area and former mills during times of extreme flood on the Merrimack River. Five days after the 1852 flood, the *Boston Daily Advertiser* said "every vestige of the old guard gates would have been carried away, and a mighty and uncontrollable river would have swept through the heart of Lowell, destroying everything in its course."

The Whistler House Museum of Art, at 243 Worthen Street in Lowell, was the home of engineer George W. Whistler, and the birthplace of his son, the renowned artist James McNeill Whistler. The museum is a charming place to visit (Fig. 8.5), as it was also the home where James B. Francis and his wife Sarah raised their six children. Worthen Street itself was named after Ezra Worthen, father of the William Worthen who served on the ASCE investigation committees with Francis. It is not often that you can see so much of a man's life more than a century after his death, but here in Lowell you can walk the streets that James Francis trod, tour the dam, canals, and locks he helped to build, visit some of the mills his work supported, and then open the door to his very home, where he ate his meals, watched their children play and grow, and laid his head to rest each night.

James B. Francis was a remarkable engineer who should be better known today for the body of his work in hydraulics and the practical engineering achievements that resulted. His work at Lowell was a triumph of the Industrial Revolution in North America. But part of Francis' legacy also rests with his role as a hydraulic engineer in the investigation of two dam failures, including the breach of the infamous South Fork Dam.

Fig. 8.5 Whistler House and Museum on Worthen Street in Lowell. Whistler sold this house to James B. Francis in 1837, and here James and Sarah raised their family. Photo by the author in 2017

8.3   William Worthen

William Ezra Worthen (Fig. 8.6) worked with James B. Francis on industrial projects at Lowell, long before they together investigated the South Fork dam. Worthen was elected President of ASCE for the year 1887, and his portrait and highlights of his career to date were published that year in the Jan. 29th issue of *Eng. News*.

Worthen was born on March 14, 1819 in Amesbury, Massachusetts to a family of long standing in that region (Bagnall 1883). Part of the family history is related here. His father, Ezra Worthen Jr., was the son of Lt. Ezra Worthen (1734–1804), who in turn was one of the sons of Ezekiel Worthen (whose father's name was probably Wathen) who helped establish the town of Amesbury in 1666, along with 35 other freemen. Amesbury is north of Boston, about 2 km south of the New Hampshire border. Ezra Worthen, Jr. became known for his skills in woodworking and machinery, and he served in a variety of jobs. That he was an innovator became clear around the year 1809 when he built the first four-wheeled wagon ever seen in that part of Massachusetts (Bagnall 1883). Worthen bought a small interest in a sawmill enterprise in 1810, and the following year acquired land for a machine shop where it is believed he manufactured carding machines. Carding is the process of

Fig. 8.6 Portrait and signature of William Worthen (source: *Eng. News*, Jan 29, 1887). Image was to appear in the Jan 22nd issue of *Eng. News*, but an accident to the finished plate, while in custody of the photo-electrotyper, delayed publication by a week

untangling, cleaning, brushing, and aligning fibers or hairs to prepare them for textile manufacture. This process is applied to cotton, wool, polyester, and other fibers, and is followed by spinning.

Worthen's early business earnings led to his future ventures in textile manufacturing in Lowell. In 1812, in partnership with three other men, he built a sizeable mill to make woolen goods (Bagnall 1883). In 1813 their business was incorporated by the state legislature as a joint stock company that raked in big profits due to the ongoing War of 1812 and the great demand for broadcloth for army uniforms. Profits slumped after the war when large volumes of imported textiles flooded the market. Worthen remained associated with the business until the spring of 1822, when he was asked to move to East Chelmsford to become superintendent of the newly built Merrimack Manufacturing Company.

East Chelmsford was a farming community that became incorporated as the city of Lowell in the 1820s. Worthen himself had previously suggested the area as having all the attributes needed to develop industry. He made this suggestion to his friend of more than 12 years, Mr. Paul Moody, who relayed the idea to Patrick Jackson and Nathan Appleton. These men carried out that vision and created the Merrimack Manufacturing Company. Ezra Worthen, Jr. served this company from June, 1822 until the very moment of his death, on June 18, 1824. He suffered from heart disease, and for several years had been afflicted by angina pectoris. According to Bagnall (1883):

> He [Ezra Worthen] often suffered paroxysms of pain, and his physician had repeatedly cautioned him to avoid excitement and overwork. This was hardly possible for a man of his temperament. At the time of his death he was superintending the labor of some men who were digging for the foundation of a building, when, seeing one of them working awkwardly and inefficiently, he seized the spade to show him how he should handle it, and as he raised the spadeful of earth, he was seized with a spasm, fell to the ground, and expired instantly.

For Ezra Worthen's role in helping to found the industrial city of Lowell, Worthen Street was named for him. Likewise, there are streets christened after Patrick Jackson, Nathan Appleton, and Paul Moody. A street in Chelmsford, southwest of Lowell, is named after James B. Francis. As of the year 1826, Ezra Worthen's widow lived north of the Merrimack River on the west side of Bridge Street, in a house situated between West Third and West Fourth Streets. A fascinating and detailed account of Lowell, its streets and residents, in the spring of that year is provided by A. B. Wright (1886). He was only about 7 years old at the time, but evidently had an extraordinary memory that was augmented by a lifetime of experiences in Lowell.

At the time of his father's death, William E. Worthen, who would one day become president of the ASCE, was not yet 5 years old. William was born at Amesbury, MA 2 years before his father began work at the Merrimack Manufacturing Company. His father had been earning a salary of $1500 per year, had been given shares in the company, and also received free housing. On his father's death there was undoubtedly an estate of some value, as William attended Harvard and graduated in 1838 at age 19. One of his classmates was James Russell Lowell, who became a renowned poet, associated with the so-called Fireside Poets, and was also an editor and diplomat. Immediately on graduation Worthen began working for George R. Baldwin, measuring the discharge rate of water used at Merrimack Mills, and was also engaged at the Jamaica Ponds water supply works in Boston.

Afterward, William Worthen came home to Lowell to work under James B. Francis doing mill and hydraulic work. Thus began William's long association with Francis. In 1840 he was employed by George Whistler on the Albany and Stockbridge Railroad, and like Francis he benefitted greatly from working for this pioneering railroad engineer. Worthen returned to Lowell to again work with James Francis on hydraulic projects and construction of the lower end of the Northern Canal, which parallels the present-day Father Morissette Blvd.

The following partial list of Worthen's engineering endeavors gives a good picture of the varied projects he engaged in before becoming President of ASCE:

Designed and built a dam and mills near the present town of Pembroke, on the Suncook River, a tributary of the Merrimack River

Designed and built a dam and mills for the Boston Manufacturing Co. on the Charles River, and served as acting superintendent of these mills and associated machine shops

Designed and built five mills at Lowell

In New York did architectural work for structures at No. 200 Broadway and the bindery of the Appleton Publishing Co.

In 1851 was supervising the machine shops and cotton mills of the Matteawan Co.

Served as engineer and later the Vice-President of the New York and New Haven Railroad

Designed and built dams across the Bronx River at West Farms

Designed floating grain docks; worked on steam-heating for buildings; and designed pumping engines for the water supplies of Cincinnati and St. Louis

Served as Sanitary Engineer for the NY Metropolitan Board of Health since its inception in 1866

Finished the Southern Boulevard in New York and built the first pumping engine at High Bridge, New York City, and many smaller such engines

Extended the water supply of Cohoes, NY and completed the waterworks of Long Island City

In 1890 and 1891, served as chief engineer of the Chicago Main Drainage Canal

Worthen was elected a member of ASCE on December 4th, 1867 and was given honorary membership on April 4th 1898. He had also been an honorary member of the Engineers' Club of Saint Louis, MO. There is no question that Worthen, like Francis, was exceptionally qualified to investigate the disaster caused by the failure of the South Fork dam. His engineering achievements are evidence of an exceptional energy and intellect. Worthen's career accomplishments reveal that he was a master of many specialties.

Worthen was also the author of many professional papers. His works in the ASCE transactions addressed the following topics: algae and purity of water supplies, back-water in streams as produced by dams; concrete sewer at Mt. Vernon, NY; crushing strength of American iron; wind pressures; Dunning's Dam; high masonry dams; hoisting apparatus of canal head-gates; improvement of sedimentary rivers; mean horsepower of a stream; Presidential address at Hotel Kaaterskill, NY; quality of water supplies; rainfall and river flow; committee report on failure of the Mill River dam; committee report on cause of failure of South Fork dam; self-purification of flowing water and influence of polluted water in disease; sewerage of Memphis; steam heating; storage and pondage of water; tests of cement; and the Holland Dikes.

Worthen published academic works that included *First Lessons in Mechanics*, *Rudimentary Drawing for Schools*, the *Cyclopaedia of Drawing* (Worthen 1896) and *A Practical Treatise on Architectural Drawing and Design* (Worthen 1862). What could better illustrate his attention to engineering detail? These works were needed to train artists and draftsmen to accurately portray engineering plans so that products of machine shops would have the requisite dimensions, and that builders of bridges, dams, and buildings could faithfully follow the engineers' physics-based designs.

This attention to detail continued until shortly before Worthen's death, when he had just finished a major rewrite of the *Cyclopaedia of Drawing*. He suffered a paralyzing stroke in December of 1896 and could not recover. William Worthen passed away on April 2, 1897. His obituary appeared in the *New York Times* (1897) the next day. He died at his apartments in New York at 436 Lexington Avenue, survived by his wife, the former Margaret Hobbs of Boston. They had no children. The funeral was held in New York at St. Bartholomew's Church (now also known as St. Barts), with later interment at Lowell, Mass. The engineering world of the late 1800's had lost one of its giants. His obituary declared that, with his passing, "the profession of civil engineering lost a man who stood at its very head. His knowledge of his profession was regarded as almost phenomenal."

A memoir of Worthen was published in the ASCE transactions a year and a half after his death (Adams et al. 1898). It was written by his friends and colleagues, Julius Adams, George S. Greene,[6] Henry Flad, Joseph Davis, and Alphonse Fteley, who said of Worthen that there was:

> …scarcely a branch of civil or mechanical engineering wherein his professional fitness has not been conspicuous in a marked degree. It was not his lot to project or carry to completion great works of internal improvement, such as challenge the admiration of the unthinking public, but in a very unobtrusive way he continually rendered that essential service toward the furthering of enterprise, without which the best conceived projects would prove abortive. To a remarkable power of rapid generalization, seemingly incompatible with painstaking accuracy, he united an almost intuitive perception of the requisite expedients of detail and design. His quickness in technical analysis, combined with the before-mentioned qualities, rendered possible the successful completion of many important works with which, owing to a forgetfulness of self, his name is scarcely associated. A retentive memory, to sift and treasure the facts in science and art which extended study had opened up to him, the tact of judicious selection and application, originality and boldness at times bordering on audacity, and a positiveness that silenced all opposition, have been the characteristic features of his long practice…
>
> Mr. Worthen possessed an overflowing vein of wit and humor; which served to temper the asperities not unusual in professional debates; this, coupled with a kindliness of disposition which could see nothing in others, to speak of, but what was commendable, led to his friends being numbered only by his acquaintance. His social relations were of the happiest kind, and his memory will long be prized by all who knew him…..

8.4   Alphonse Fteley

A description of the life and career of Alphonse Fteley appeared in *The Engineering Record*, issue dated June 20, 1903 (p 673), along with his portrait. He had died on June 11th, just prior to ASCEs annual convention held in Ashville, NC. He was survived by his daughter Estelle (1874–1943) but not by his wife Elise (nee Maurer, b. 1846 in Switz.), who had passed 5 years earlier at age 52. Private announcement of his death was given at that Ashville meeting. Word rapidly spread of the loss of

---

[6] George Sears Greene was one of the original founders of the American Society of Civil Engineers and Architects. He gained fame during the Civil War commanding troops in many battles. At Antietam his small division deeply penetrated Stonewall Jackson's lines and with little support held their ground for four hours. At Gettysburg his lone brigade defended Culps Hill, driving off repeated attacks by a Confederate division.

this highly respected engineer who had served ASCE as a Vice President and later as President in 1898. The annals of *Eng. News* contain many references to him, commenting on papers and issues before the Society, including numerous articles about his work on the Croton Aqueduct. *Eng. News* on Jan. 20, 1898 summarized his career up to that time. Both that periodical and *The Engineering Record* were drawn upon in the following notes.

Fteley was born in Paris, France on April 10th, 1837. He was educated at the Ecole Polytechnique which at that time was housed in central Paris. That institution had been established in 1794 during the time of the French Revolution, and continues today as a top-flight engineering school. He worked for 6 years in general practice with several engineering firms in Europe, focusing at that time on mill engineering. Fteley emigrated to the U.S. in 1865 and was employed as a mechanical draftsman at a machine works in New York, mainly on projects related to construction of steamboat engines. In 1866 he began his association with William Worthen, whom he assisted on various water supply and hydraulic power projects.

From 1870 to 1873 Fteley ran his own engineering practice, concentrating on highways, bridges, and waterworks. In 1873 he relocated to Boston to supervise the new waterworks that would provide Boston with water from the Sudbury River. He did this work under Mr. Joseph Davis, who was City Engineer for Boston. The Sudbury aqueduct was eventually built as part of this effort, and during 1875 to 1880 Fteley was occupied with building that aqueduct, associated reservoirs, and other engineering projects related to the water supply. He was elected to membership in the ASCE on Jan. 5, 1876.

When Joseph Davis resigned in the spring of 1880, Fteley was promoted as Chief Assistant City Engineer of Boston, which duties he performed until 1884. During this time Fteley did important work with his assistant, F. P. Stearns, bringing them both to the attention of the broader engineering world. They ran a series of water discharge experiments related to weirs, and their results were presented to the ASCE and published in the Society's transactions (Fteley and Stearns 1883). ASCE awarded the Norman Medal jointly to these men in 1882 for this substantial work. That medal continues to be awarded today. Instituted in 1872 by George H. Norman, it recognizes a professional paper that makes an important contribution to either research or practical aspects of engineering. The hydraulic work by Fteley and Stearns is discussed in some detail in a U.S. Geological Survey paper on hydrographic investigations (Horton 1906).

While serving as chief assistant to the Boston city engineer, Fteley worked on the design of various municipal projects; including key parts of the main drainage, water supply and park systems. His association with Worthen continued, as shown by the fact that both served on a committee appointed by the Harvard board of overseers (Harvard Univ. 1881). Worthen was a Harvard alum and likely lobbied for Fteley's inclusion. During the 1881–82 academic year their committee visited the Lawrence scientific school, the Bussey institution of agriculture, the Peabody museum of American archaeology and ethnology, and the museum of comparative zoology.

In 1884 Fteley relocated to New York to become the Principal Assistant Engineer to the Croton Aqueduct Commission of New York, where he played a prominent role in the general design of this large urban water supply system. He resigned from this position 2 years later and entered private practice, but remained as Consulting Engineer for the Commission. He became Chief Engineer of the Commission in 1888. Fteley was one of the two Vice Presidents of ASCE elected in 1889 (the other was Elmer L. Corthell of Chicago). Within a week of the Johnstown flood he was appointed to serve on the Committee to investigate the disaster, which included William Worthen, his friend of long standing.

Due to his highly respected skills and extensive knowledge, Alphonse Fteley was much sought as a consulting engineer and as an expert witness in court cases. He was a member of the technical advisory committee of the New Panama Canal Company and was consulting engineer to Newark during construction of the East Jersey Water Company's works. *The Engineering Record* (1903) listed some of his other consulting projects:

Constructing engineer for the completion of the Southern Boulevard in New York, from Third Avenue to Jerome Park, and for several minor water supply systems.

Reported on the lining of the Hoosac Tunnel; on additional water supplies for St. John, N. B., Albany, N. Y., Cincinnati, O., Brooklyn, N. Y., Rochester, N. Y., Newton, Mass., Cambridge, Mass., and other places; also reported on reservoirs and dams for a number of cities.

Examined and reported on sewerage systems for Brooklyn, N. Y., Hoboken and Newark, N. J., and on the purification of the valley of the Passaic River. Fteley had been appointed a commissioner of the Passaic Drainage Commission, which provided a report to the New Jersey legislature.

Reported on the accident to the Brooklyn Aqueduct, in 1891, and regarding the advisability and cost of repairing the aqueduct tunnel in Washington, D. C.

Consulted for the first Rapid Transit Commission of Boston, for the Rapid Transit Railroad Commission of New York, and also to the Long Island Railway Company, on the proposed tunnel under the East River.

Served as a consultant for the construction of supplementary water supplies for Rochester, N. Y., and did similar work for the Service of the Metropolitan Water Board of Massachusetts.

Fteley published papers in ASCE's transactions on the following topics, four of which were jointly authored with his mentor, William Worthen: covered reservoirs; algae and the purity of water supplies; cements, mortars and concretes; experiments on water flow; the Oakley Arch; friction and loss of water in mains; high masonry dams (Fteley 1889); main relief sewer of Brooklyn; ASCE presidential address in 1898; proportional water meter; rainfall, stream flow, and storage; experiments with dynamite on an ocean bar; report of the committee on the failure of the South Fork dam; stability of bench marks; storage and pondage of water; standard levee sections; subaqueous foundations; the Sweetwater Dam; the flow of the Sudbury River, 1875–79; the nozzle as an accurate water meter; the Shone hydro-pneumatic system of sewerage; the water works of Denver; and water power with high pressures and wrought-iron pipe.

In appreciation of his work in and around Boston, Fteley was made an honorary member of the New England Water Works Association. A Bostonian friend of many years said of him:

> He made a host of friends both in and out of the engineering profession and established firmly in this vicinity his reputation as an eminent engineer. It was characteristic of the man that those in his employ not only had the highest respect for his attainments, but they became deeply attached to him as a friend. He was thoroughly liked even by the contractors from whom he required first-class work. That an absence of twenty years has not diminished the attachment of those who knew him well is evidenced by the eagerness with which news regarding his health has been received in recent years.

Fteley became Chief Engineer of the Croton Aqueduct Commission in 1888 and stayed in this position for 11 years. He was elected President of the ASCE in 1898. That year the annual convention was held in Detroit, MI, where Fteley (1898) gave a detailed address on the state of engineering in the U.S. The text of his speech covered more than 20 pages in the June issue of ASCE's *transactions* (1898). The following year he retired from engineering work due to health problems involving a complication of illnesses. The June 20, 1903 memorial article published by *The Engineering Record* had unusual comments about Fteley's work with the Croton Aqueduct Commission. These comments have some bearing on his philosophy in dealing with professional disagreement with authorities:

> The facts concerning the difficulties of his [Fteley's] work have never been fully published, and this is no place in which to discuss them. It is sufficient to say that not infrequently his recommendations were overruled by the [Croton] Commission for reasons of public policy, the force of which he recognized. The great dam now approaching completion is erected on a site which Mr. Fteley never approved, and other less striking but very important instances might be mentioned. The Commission received his loyal support at all times, however, even when it ordered work conducted on lines not the most desirable from an engineering standpoint.

The comments about Fteley are relevant to his participation in the ASCE investigation of the South Fork dam and the Johnstown flood of 1889. To what extent would he have objected if the final report disagreed with his own conclusions? Fteley was a Vice President while the other three committee members were the standing president of ASCE and two former presidents. It is a matter of record that he left on an overseas trip soon after the investigation committee returned from Johnstown. *Eng. News* announced in its "personal" section that he sailed for Europe on July 31 to begin a month's vacation. In his absence, Deputy Chief Engineer George Rice was in charge at the New Croton Aqueduct. Fteley's trip raises a question about how much input he had in preparing the initial draft of the investigation report. The report was completed in January of 1890, so there was plenty of time after his return for Fteley to help write and approve the final conclusions. He had spoken to the ASCE membership at the June, 1889 meeting, stressing that the "truth of this matter should be known, as it may show that other dams are unsafe…" (*Eng. News* 1889d, p 603). Because he said this at the ASCE meeting, it is doubtful that Fteley would readily have agreed to delay publication of the investigation report until 2 years after the Johnstown disaster. But that delay did happen.

After Fteley died, *The Engineering Record* (1903) gave testimony to his exceptional character:

> The number of small works on which he gave advice was legion, and there are many young engineers who will bear testimony to his patience in hearing their troubles, his skill in solving them and his frequent generous refusal to accept any fee for his valuable advice. A man of deep and broad intellectual attainments, a keen judge of character, gifted with a charming manner, and possessed by a sterling honesty against which suspicion never even pointed, his death is a great loss to the community he served so faithfully.

Some final notes about Fteley come from Johnstown itself. Years after the 1889 flood, plans were laid to build a large dam and reservoir on Quemahoning Creek, a tributary of the Stonycreek River above Johnstown. The impetus for this dam was the need for a reliable water supply for the Cambria Steel Company, which had a bearing on existing and future jobs in Johnstown. But the notion of building that big dam above the town understandably raised fear and suspicion among residents of the rebuilt city, survivors of the 1889 calamity. That was especially true when planners would not divulge specific details about the dam.

A reporter with the *Daily Tribune - Johnstown* (1900) interviewed engineer John Birkinbine who said, "I don't blame the people a bit for feeling unease about a new reservoir above them, after the experience they had with the South Fork Dam." He said the dam would be built to "…be as safe as human agency can make it." But he also said "It would be manifestly unfair for me to talk about this matter at this time." He had been retained by the Manufacturers Water Company (a subsidiary of Cambria Steel Company) to take charge of the planning and construction, and was assisted by a Mr. Gowan. Birkinbine told the reporter that he and Gowan had been "going over the ground and securing data and information on the subject." He then went on to say,

> Mr. Gowan and myself will then go to New York and lay the results of our labors and research before Alphonse Fteley, who is justly regarded as the highest authority in dam construction in the country. He is the designer, and until his health broke down was the engineer in charge of the construction of the immense Croton reservoir above New York City, having a capacity of 30,000,000,000 gallons. Mr. Gowan, who is with me now, has been his assistant in the Croton work and is himself an authority on the subject. His paper on the Croton Dam read before the American Society of Civil Engineers was a notable work and was so regarded in scientific circles…When we return to New York we will go into consultation with Mr. Fteley and there the problem will be gone over carefully, for this is a matter which will not be lightly considered.

By the time of that interview, James B. Francis, William Worthen, and Max Becker were all dead. Fteley himself was ill and would die in less than 3 years. But his review of the plans for the Quemahoning Dam would eventually bear fruit. *Eng. News* (Aug. 3, 1905) reported that work was to begin on August 1, 1905. However, the *Bull. of the American Iron and Steel Assoc.* (1912) wrote that work actually started in 1908 and the dam would be complete by October 1913. The article noted that the breast of the dam had a concrete core which extends down to solid rock. The construction of the dam was supervised by Charles Curtis, from the engineering force of the Cambria Steel Company.

The consulting engineer for the Quemahoning Dam was F. P. Stearns, who helped build the Gatun Dam on the Panama Canal. He was also the former assistant of Fteley, with whom he was honored by ASCE's Norman Medal in 1882.

8.5 Max Becker – Railroad Man

A detailed summary of Max Becker's career, along with his portrait (Fig. 8.7), was published by *Eng. News* (1889c) following his election as President of the ASCE. Some of the material that follows is abstracted from that source and other issues of *Eng. News*.

Max Joseph Becker was born on June 1st, 1827 at Koblenz in Prussia, now a part of western Germany. This city lies on both sides of the Rhine River where it is joined by the Moselle River. He attended a college in Koblenz where his father taught. In 1846 he began a volunteer apprenticeship on the staff of the chief engineer of the Department of Public Works. The engineer's work related to the locks and canals on the Luhn River. While so employed, in 1847 Becker accepted a job offer to help survey the Cologne-Minden Railroad. That year he worked on several survey parties between Dusseldorf and Bielefeld. He became assistant engineer in charge of construction of a division, in 1848, and supervised the building of a stone arch bridge that crossed the Lippe River.

Fig. 8.7 Portrait and signature of Max J. Becker (Source: *Eng. News* 1889c)

Becker's early and extensive experience with the new railroad technology, promising rapid growth, would have eventually served him well in Germany. However, 1848 was a year of revolutions in Europe, with widespread political unrest involving a series of revolts against monarchies by their common people. A Palermo revolt began in January in Sicily, and other unrest followed in France, Hungary, Germany, Prussia, Italy, and Ireland. Most of the revolts failed in their purposes, although the February uprising in France ended the monarchy of King Louis-Philippe. The French Second Republic was born.

Max Becker participated with others in what may have amounted to armed rebellion, but that revolt entirely failed in 1849 and those who could travel fled and were exiled in Switzerland. He managed to immigrate to the United States in May of 1850. Others who immigrated for similar reasons included the well-known Franz Sigel, who became a Union major general during the Civil War, and Carl Schurz who was also a Union general and later served in high political offices, including U.S. Senator from Missouri. Becker had little success finding gainful employment until the spring of 1851, when he got work surveying the geology and topography of parts of Connecticut. That December he went to Ohio to work on the Steubenville and Indiana Railroad. His supervisor was Chief Engineer Jacob Blickensderfer, and Becker was engaged as a draftsman, mapmaker, and assistant engineer for railroad construction until that line was finished in 1855.

Becker served as an engineer for the State of Ohio, on the Board of Public Works, from 1856 to 1858, and returned to railroad work during 1859 to 1861 on the Marietta and Cincinnati Railroad. In 1862 he was again employed by the Pittsburgh, Cincinnati and St. Louis Railway (the former Steubenville and Indiana RR), where he supervised building the first railroad bridge to span the Ohio River, at Steubenville.

By 1863 Max Becker was back with the Marietta and Cincinnati Railroad where he supervised construction of its western portion from Loveland to Cincinnati, during the late Civil War years of 1864 to 1865. He became Chief Engineer of the Pittsburgh, Cincinnati and St. Louis Railway in 1867, a position he still held at the time of his election to be President of ASCE. He had been elected a member of ASCE on August 7, 1872, and joined the Engineers' Society of Western Pennsylvania on January 6, 1880.

Baer (2005) has done extensive work in reviewing and documenting the history of the Pennsylvania Railroad. In April, 1884 the Chicago, St. Louis and Pittsburgh Railroad was organized, its president being George Roberts with Max Becker as Chief Engineer. In November, 1887 the Real Estate Department of the PC&StL/CStL&P (Pittsburgh, Cincinnati, & St. Louis/Chicago, St. Louis, & Paducah) was formed under Chief Engineer Becker, who also became Real Estate Agent on the manager's staff (Baer 2013).

Max Becker's office moved from Columbus to Pittsburgh on January 1, 1888 (Baer 2011). Before this he would have had to closely coordinate with Robert Pitcairn's PRR office for the movement of passengers and freight across their respective lines. But now Becker would be in the same city as Pitcairn and some of these dealings would be face-to-face. Very soon Becker would be part of ASCE's investigation of the South Fork dam breach, where Pitcairn had been a member of the prestigious Club on the lake.

At the start of 1889, when he became President of ASCE, Becker was supervising 1600 miles of railroad lines extending from Pittsburgh in the east, westward as far as Illinois. He was responsible for four large bridges over the Monongahela and Ohio Rivers. Also while President, Becker had become a Director of the Engineers' Society of Western Pennsylvania and had just been reappointed as City Engineer of Austin, Minnesota, located south of Minneapolis.

One of the bridges for which Becker was chief engineer was entered in the National Register of Historic Places in April, 2001. The Newport and Cincinnati Bridge spans the Ohio River between those cities. It was erected during August 1896 to February 1897 by the Newport and Cincinnati Bridge Co. It is also known as the Louisville and Nashville Bridge, and is unique because it is the only bridge designed and built for dual railroad and highway use in Kentucky. It includes a pedestrian walkway.

Becker published papers on the following topics in ASCE's transactions: cause and prevention of decay of building stone; cylindrical wheels and flat topped rails for railways; the Oakley Arch; destruction of rails by excessive weights; comparison of English and American railroads; ASCE presidential address in 1889; proper relation of the sections of railway wheels and rails; properties of steel, and its use in structures and heavy guns; reconstruction of the Cork Run [railway] Tunnel; relation of wheels to "frog points" and to guard rails; the incline plane railroad at Madison, Indiana; and the report on the failure of the South Fork Dam. Becker was not a hydraulic engineer, so the question should be posed – why was he on that investigation committee?

8.5.1   ASCE Election Controversy

Eng. News chronicled a controversy that arose over the 1889 election of ASCE's President. Official ballots had been distributed to the members, with names for offices chosen by a Nominating Committee that had been appointed at the Milwaukee Convention in 1888. Shortly after the ballots were sent out, "unknown persons" distributed alternative ballots under the title "Irregular Ticket for Officers of the American Society of Civil Engineers." In the alternative ballots, Max Becker's name had been removed and the name Thomas C. Clarke substituted. Names for other offices were unchanged. This was a clear attempt to circumvent the nominating committee and prevent Becker from becoming President.

In response, the following letter from an *anonymous* member of the society appeared in *Eng. News* (1889a p 14). This was in advance of the election to be held at the annual meeting 11 days later. The letter is reproduced here as it sheds light on how elections and the office of President of the ASCE were perceived at that time:

The Opposition Ticket for President of the - Am. Soc. C. E.

Chicago, Ill. Dec. 31, 1888.

EDITOR ENGINEERING NEWS:

It is much to be deplored that the next election of a President of the Am. Soc. C. E. is to be made the occasion of personal gratification by the presentation of another name in the eleventh hour. The regular Nomination Committee of the Society nominated for President, Mr. MAX J. BECKER, Chief Engineer of the Pittsburg, Cincinnati & St. Louis R. R., an eminent and old engineer of the highest character, both professionally and socially, and holding a high position similar to the one held by one of our Ex-Presidents, Mr. Don J. WHITTEMORE. Mr. BECKER, as is well known, accepted the nomination only after he was distinctly assured of its unanimity by the Committee. Personally, I know he did not seek or expect the honor. It has become understood in the Society that the position of President is a mark of honor to be bestowed as a rule only on older and distinguished members, and the usual short duration of the office makes it possible to reach successively nearly all those deserving the honor. The election after the nomination by the Nominating Committee is understood to be a mere matter of form. The duties are not so arduous that they cannot be fulfilled by a President living at a distance from New York, and there was no objection heretofore to selecting candidates from a distance, if deemed deserving of the honor. Do all the distinguished engineers live in or near New York? No one would claim that much I trust. The presentation of the name of Mr. T. C. CLARKE of New York for the presidency of our Society, with the rest of the ticket unchanged, and at the last moment, is a somewhat remarkable proceeding, particularly as it is not endorsed by the signatures of any committee or otherwise, and it does not appear who is responsible for this electioneering manoeuvre. Inference may easily be drawn which I do not care to specify. Mr. CLARKE is a very estimable and well known gentleman and a shrewd business man, to whom the honor may not be without some business value; while such an aspect of the candidacy is entirely absent in a man in Mr. BECKER'S responsible position, trusted and honored by his company and beloved by his subordinates. At least so the matter must appear to most of those having the pleasure of personally knowing both the candidates. For the good and dignity of the Society it would be well if such rivalries, particularly those thrust on the members at the last moment, could be avoided, and a change in the mode of nominating and electing our president seems highly desirable to that end, though it may not be easy to say precisely what the change should be.

A MEMBER of the Am. Soc. C. E.

This letter suggests that some members were troubled about having an ASCE president who lived outside of New York, as though this would hamper his conduct of the office. However, the primary concern was revealed (see underlined text) in a second article, on p. 16 of the same Jan 5th issue of *Eng. News*:

Inasmuch as an opposition ticket has been sent out by unknown persons, in which the nominee of the regular ticket for President is ignored, we submit the following remarks concerning the regular ticket: The Nominating Committee was appointed at the Milwaukee Convention, with a special view to nominate a ticket that would represent all sections of the country. The Committee met in the city of New York, and after a full and free discussion, unanimous nominated the ticket, which has been regularly issued by the Board of Directors, bearing the name of MAX J. BECKER, thereon for President. Under the pretext that he is not an American-born citizen, it has been undertaken to defeat his election by sending out an irregular ticket, having thereon all the names that are on the regular ticket, except that of Mr. Becker. Mr. Becker is an eminent and successful engineer of the highest standing, socially and professionally, well qualified to perform the duties of President of the Society, and having been nominated by the duly appointed Committee, should receive the suffrages of the members, unless there is some good reason for withholding them. The undersigned members, feeling very strongly that an injustice is being attempted, as also an injury to the Society, unite in requesting you to vote the regular ticket.

Charles Paine, a past President of ASCE, and 16 other members signed their names to the above. It is a credit to them - it appears Max Becker was being discriminated against as an immigrant. Being at the center of a whirlwind, Mr. Clarke felt obliged to comment on all of this, especially after he was asked in writing to withdraw from the election. *Eng. News* (1889b) published his note on Jan. 12th, just days before the election.

A Card from Mr. Thos, C. Clarke
No. 1 Broadway, NEW YORK, Jan. 9, 1889.

To The Editor of Engineering News:

I have received a letter from a member of the American Society of Civil Engineers, asking me to withdraw my name as candidate for the office of its President. I shall not do so, and wish to publicly give my reason for this course. I believe it to be for the best interests of the Society that there should be a contested election. Some members holding this opinion issued a second ticket. Had they done so before Dec. 1, the Society would have, under Clause 25 of its by-laws, borne the expense. They preferred to do that themselves. If this be "irregularity"—let the most be made of it. But the real point, which some members try to make is, that to vote for a second ticket casts a slur upon the first. If it were admitted that nobody could bring forward a second ticket, either under Clause 25, or otherwise, then it would follow that the Nominating Committee and not the Society itself elected the officers. That is the strict logical consequence of members being not allowed to vote upon a second ticket. As to the personal charges, I do not believe them, and shall not notice them. There is no odium in not being elected. It simply means that more members prefer one man than others do. If it is Mr. BECKER, my heart will not be broken, and I hope he feels as I do. A still more important reason why I do not withdraw is, that I hope this year's contested election will bring about the changes in mode of nominating candidates desired by many of our members.

Your obedient servant,
THOMAS C. CLARKE.

The election was close. Max Becker defeated Clarke, who garnered 48% of the vote, 282 ballots versus 311 cast for Becker. If indeed an anti-immigrant bias against Mr. Becker was the main cause of the dispute within ASCE, then I must commend the members of that society for rejecting that bias, at that time in history, however close the margin may have been. After all, one of the most preeminent and early members, James B. Francis, was himself an immigrant from England and a self-made man of little formal education. And no doubt many other members, including Alphonse Fteley, had also immigrated to the U.S. with varying levels of education. But the desire for rational change in how the Society's elections would be handled, including nominations and ballots, was now out in the open, as described in a two-column editorial of *Eng. News* (1889a).

Thomas Curtis Clarke was certainly qualified to be President of the ASCE, and was himself elected to that office 7 years later in 1896. Born in 1827 in Newton, Massachusetts, he was a Harvard graduate. His engineering works on buildings, railroads, and especially large railway bridges, along with his many professional papers, were well known in Canada, the U.S., and in Britain. Clarke died in New York, where he had lived at 146 East 38th St., but was interred in Port Hope,

Ontario where he had been married and raised his family. Clarke's obituary appeared in the *New York Times* (1901) on June 17. He was interred at St. Johns Cemetery.

Becker's career peaked with his position as Chief Engineer of the entire system of the Pittsburgh, Cincinnati, Chicago and St. Louis Railway. He took on additional responsibilities on March 25th, 1890, during the time the report on the South Fork dam was still being kept under wraps. The Board of the Ohio Valley Railway named him to be Chief Engineer, replacing Isaiah Linton who had resigned. Linten died the following year.

Due to an illness that afflicted him, in January, 1895 the railway Board of Directors designated Becker as Consulting Engineer and Real Estate Agent, to reduce some of his duties. Max Joseph Becker died at age 63 on August 23rd, 1896. He was survived by his wife and two daughters. His passport application from May 25, 1894 shows his wife's name was Ellen A. Becker, born in Knox County, Ohio. Their daughters Minnie L. and Nellie M. were 21 and 28 years old, respectively. At that time Becker and his wife lived at 5525 Ellsworth Avenue, Pittsburgh.

Becker's career was described in an obituary drafted by ASCE members William Metcalf, Thomas Johnson, and Samuel Rea, and published in the *ASCE transactions* (1897 p 555–557). They obliquely mentioned the questions that arose during the 1889 election when he became President. Of this and of Becker personally, they wrote:

> At the time of his election his personal acquaintance in the Society was not very large, and some questions were asked; before his term had expired he had won all hearts, as he always did wherever he came into touch with people. Mr. Becker was more than an able engineer, he was a man of force, a gentleman, genial and witty, kind and sympathetic, always ready to lend a helping hand. A man without malice or guile, he had no time for animosities, and his memory will ever remain in the minds of all who knew him.

Looking back, a part of Max Becker's legacy must be his role in the investigation of the Johnstown flood and the failure of the South Fork Dam. James B. Francis said it was Becker's decision to delay publication of the investigation report for 2 years. That delay was not ethical from the standpoint of public safety. Becker himself, in his Presidential address to the Society just weeks after the flood, had stated that the data collected by visiting the dam location "…will enable the committee to submit to the society in due time a comprehensive statement of the conditions and circumstances which have induced and contributed to this most disastrous failure." No one then knew just then, perhaps not even Becker, that "due time" would be 2 years.

It would appear that someone applied strong pressure to Becker to seal and delay the South Fork report. One person with great authority over Becker was Robert Pitcairn, a senior executive of the Pennsylvania Railroad, a company that had a controlling stock ownership over the railroad where Becker was Chief Engineer. All movements of trains on Becker's railroad moving to or from Pittsburgh had to be coordinated through Pitcairn's office. Pitcairn was a boyhood and lifelong friend of the powerful industrialist, Andrew Carnegie, and both were members of the South Fork Fishing & Hunting Club and the elite Duquesne Club of Pittsburgh.

References

Adams JW, Greene GS, Flad H, Davis JP, Fteley A (1898) "Memoirs of deceased members" William Worthen. ASCE Trans 24(7):709–711 Sep 1898

ASCE (1897) Memoir of Max Joseph Becker. Am Soc Civil Eng Trans 37(June 1897):555–557

ASCE (1898) Trans. Amer Soc Civil Eng, June

ASCE (1901) Memoirs of James B. Francis. Am Soc Civil Eng Trans 45(June 1901):627–628

Baer CT (2005) Salvaging history. Railroad History, Spring-Summer 2005, 76–87

Baer CT (2011) A general chronology of the PA Railroad Co., its predecessors and successors, and its historical context, 1888. PA Technical and Historical Society. Link: http://www.prrths.com/newprr_files/Hagley/PRR1888.pdf

Baer CT (2013) A general chronology of the PA Railroad Co., its predecessors and successors, and its historical context, 1887. PA Technical and Historical Society. Link: http://www.prrths.com/newprr_files/Hagley/PRR1887.pdf

Bagnall WB (1883) Sketch of the life of Ezra Worthen. In: Contributions of the old residents' historical association, vol III, Lowell No. 1, Sep 1884

Bull. of the American Iron and Steel Assoc. (1912) Jan. 1 issue

Eng. News (1887a) Engineering News, Jun 25

Eng. News (1887b) Engineering News, Jan 29

Eng. News (1887c) Engineering Neews, Jan 1

Eng. News (1887d) James B. Francis (with portrait). *Engineering News*, Personal Section, Vol. XVII, Jan. 1, 1887, p 14

Eng. News (1887e) Engineering News, Feb 5, p 98

Eng. News (1889a) Engineering News, Jan 5, p 14

Eng. News (1889b) Engineering News, Jan 12

Eng. News (1889c) *Engineering News*, Feb 9 [Becker portrait between p 120 and 121]

Eng. News (1889d) Engineering News, Jun 29, p 603

Eng. News (1898) Engineering News, Jan 20

Eng. News (1899) Engineering News, Jun 22

Eng. News (1905) Engineering News, Aug 3

Engineering Record (1903) Jun 20, p 673

Fitz Gerald D, Davis JP, Freeman JR (1894) James Bicheno Francis: A Memoir. Journal of the Assoc. of Engineering Societies, Vol XIII, No. 1, Jan 1894, p 1–9

Francis JB (1855) Hydraulic Experiments being a Selection from Experiments on Hydraulic Motors, on the Flow of Water over Weirs, and in Canals of Uniform Rectangular Section and of Short Length. Little, Brown, & Company, Boston, 156 p

Francis JB (1868) Lowell hydraulic experiments. Being a selection from experiments on hydraulic motors, on the flow of water over Weirs, in open canals of uniform rectangular section, and through submerged orifices and diverging tubes. 2nd D. Van Nostrand, 192 Broadway, NY, 251 p

Francis JB (1886) Great freshets in Merrimack River. In: *Contributions of the Old Residents' Historical Association*, vol III., No. 3, Aug 1886, Lowell, pp 252–257

Francis JB (1887) Kirk Boott and his experience in the British Army. In: *Contributions of the old residents' historical association*, vol III, Lowell No. 4, Jul 1887, p 325–333

Francis JB, Ellis TG, Worthen WE (1874) The failure of the dam on Mill River. *ASCE Trans.* vol III, no. 4, p. 118–122. Retrieved Apr 3 2018. Available at: https://archive.org/stream/transactionsofamciven03amer#page/118/mode/2up

Francis JB, Worthen WE, Becker MJ, Fteley A (1891). Report of the committee on the cause of the failure of the South Fork dam. *ASCE Trans* XXIV, 431–469

Frizell JP (1894) Reminiscences of James B. Francis. *Eng. News*. Jul 12, 1894, p 27–30

Fteley A (1889) Researches concerning the design and construction of High Masonry Dams, in view of the proposed building of Quaker Bridge Dam. Taylor, NY

Fteley A (1898) Address at the annual convention at Detroit, Michigan, July 26, 1898. *ASCE Trans.* Vol XXXIX, No. 832, Jun 1898, p 665–685

Fteley A, Stearns FP (1883) Description of some experiments on the flow of water made during the construction of works for conveying the water of Sudbury River to Boston. *Trans. ASCE* 12(1): 1–36; 12(2): 37–80; 12(3): 81–118

Harvard University (1881) *The Harvard University Catalog*, 1881–1882. *Univ. Press*, published for the university by Charles Sever

Horton RE (1906) Weir experiments, coefficients, and formulas. *USGS Water-Supply and Irrigation Paper* No. 150, Govt. Printing Office, Wash., DC, 189 p

Lewis BJ, Cimbala JM, Wouden AM (2014) Major historical developments in the design of water wheels and Francis hydroturbines. IOP Conf. Series: Earth and Environmental Sci. 22, doi:https://doi.org/10.1088/1755-1315/22/1/012020

McCullough D (1968) The Johnstown Flood. Simon & Schuster, New York, 302 p

Mills H (1892) James Bicheno Francis. Pamphlet, read to the Corp. Of the [Massachusetts] Institute of Technology (Cambridge, MA, Dec 14, 1892), 8 p. [Lawrence, MA, Dec 11, 1892]

NY Times (1897) "William E. Worthen Dead," *The New York Times*, Apr 3, 1897

NY Times (1900) "Col. John G. Parke" (obituary), *The New York Times*, Dec 18, 1900

NY Times (1901) "Thomas Curtis Clarke Dead," *The New York Times*, Was One of the Most Widely Known Civil Engineers in America, Jun 17

Penn (1883) University of Pennsylvania, Catalogue and Announcements, 1882–1883, Philadelphia, 116 p

Penn (1885) University of Pennsylvania, Catalogue and Announcements, 1884–85, Philadelphia, 129 p

Penn (1907) A Twenty-Year History of the Class of Eighty-Seven. University of Pennsylvania, Philadelphia

Penn (1922) Five and Thirty – The Further History of the Class of Eighty-Seven, University of Pennsylvania, for the Ten Year Period 1912–1922, Philadelphia

Pennsylvania Gazette (1933) Article about the passing of former student John Grubb Parke, Jr. Issue dated Nov 1, University of Pennsylvania

The Daily Tribune - Johnstown (1900) "Experts on Dam Question" Saturday, Friday, Aug 18, 1900, p 1

The Engineering Record (1903) Alphonse Fteley (Remembrance), vol XLVII, Jan-Jun (June 20, 1903), p 673

Unrau HD (1980) historic Structure Report: The South Fork dam historical data, Johnstown Flood national memorial, Pennsylvania. U.S. Department of the interior, national park service, p 242, Package no. 124

Worthen WE (1862) A practical treatise on architectural drawing and design. D. Appleton & Co., NY, 312 p

Worthen WE (1893) James Bicheno Francis – Biographical Notice. In: Proc. of the American Academy of Arts and Sciences, New Series XX, Boston, Mass., p. 333–340

Worthen WE (1896) Appleton's Cyclopedia of Technical Drawing. D Appleton & Co, NY, 923

Wright AB (1886) Lowell in 1826. In: *Contributions of the Old Residents' Historical Association*, Lowell, MA. Vol III. No. 4, July 1887. Also published in the *Lowell Courier*, Apr 24, 1886

Chapter 9
Analysis of the South Fork Dam and the Former Lake Conemaugh

Francis et al. (1891) claimed that the South Fork dam, even as originally constructed, would have overtopped and failed on May 31, 1889. That was an extraordinary conclusion, and one that was certainly good news for the Club members. The leading engineering society in the U.S. was essentially exonerating the Club, in that all of their repairs and alterations of the dam should not be blamed. The implication was that the State had originally built a flawed dam. The failure of the Club to involve qualified engineers would therefore have been of no consequence, and that had important legal implications. And for the future image and legacy of the Club, posterity could forever blame Mother Nature for the disaster. But it turns out the claim that the dam would have failed anyway was not correct.

And what did the investigation report say about the Club's most radical change, lowering the crest of the dam? The report concluded, "…we find that the embankment would have been overflowed and the breach formed if the [Club's] changes had not been made. It occurred a little earlier in the day on account of the changes, but we think the result would have been equally disastrous, and possibly even more so…." (Francis et al. 1891 p 456).

That little phrase at the end, "…and possibly even more so…", suggested that if the dam had been kept at its design height, even more water would have been impounded and eventually released in a more destructive flood. In other words, in lowering the dam the Club had possibly done a "good deed" by reducing the severity of the inevitable flood. But in fact, lowering the dam allowed it to be overtopped and was the principal cause of the dam breach.

9.1 Research Approach

These various claims by Francis et al. (1891) can be analyzed using scientific methods and by examining details and assumptions in their report. Would the dam have survived the 1889 flood had proper repairs been made by the Club? Uldis Kaktins,

© Springer International Publishing AG, part of Springer Nature 2019
N. M. Coleman, *Johnstown's Flood of 1889*,
https://doi.org/10.1007/978-3-319-95216-1_9

Stephanie Wojno, and I found the answer is "yes." Our analysis (Coleman et al. 2016) was supported by a careful review of observations made at the dam and in the region on the day of the flood. We also depended on data recorded by Francis et al. and information documented by other engineers who visited the site. To reconstruct aspects of the dam breach on that terrible day, we developed a series of key objectives:

1. using data from historical reports,[1] determine the elevation of the lake surface (in the 1889 reference frame) at the moment the dam breached;
2. determine locations at the dam where the elevation has not changed significantly since 1889;
3. using GPS (global positioning system) and LiDAR techniques, establish "benchmarks" at the site and survey the present dam remnants and spillways to compare with the 1889 survey by Francis et al. (1891);
4. reconcile the difference between the 1889 elevations and the present-day GPS elevations, using common surface points where elevations have been relatively constant;
5. convert the 1889 lake surface elevation into a modern GPS/Lidar elevation;
6. using the derived GPS elevation for the former lake surface, prepare a storage-elevation curve for Lake Conemaugh;
7. develop rating curves to evaluate the safe discharge capacity for the dam as originally built (with a higher crest, two spillways, and 5 discharge pipes);
8. compare the discharge capacity of the original dam with that of the dam as modified by the Club;
9. using the storage-elevation curve, spillway rating curves, and reports of lake rise on the day of the flood, evaluate the range of inflows to Lake Conemaugh before the dam breach;
10. evaluate the time of concentration for the South Fork watershed, and estimate the time to peak discharge based on available reports from witnesses;
11. determine the geometry of the dam breach using survey information obtained by Francis et al. (1891) two weeks after the flood;
12. model the dam breach outflows using several scenarios of breach formation; prepare hydrographs of discharge rate and declining lake level, with time zero representing the moment of general dam failure;
13. and illustrate the much longer time needed to overtop the dam had it been kept at its design height, even if extreme (and unrealistic) continuing inflows to the lake were maintained on the day of the flood.

The completion of the above objectives led to our paper in the journal *Heliyon* (Coleman et al. 2016). Analyses and results are given in this chapter and Chaps. 10

---

[1] Our research relies on many nineteenth century publications and more recent work that document historic data using English units rather than SI units. For most calculations we use the always preferable SI units, but where we highly depend on old data sources we report the original English units. We believe this will help confirm our appropriate use of the nineteenth century data and will aid future workers who may further study the Eastern Dam.

and 11. Objectives 1 through 4 were the most challenging and time consuming, but the work progressed rapidly after we reconciled the differences between the 1889 elevations and modern GPS data. Appendix 2 describes our process for achieving this. We are confident that the difference between the 1889 and present-day elevations is in the range from 6.1 to 6.3 ft., with 6.2 ft. (~1.9 m) being our best estimate. Given this information, profiles of the dam remnants were prepared.

9.2 Profiles of the Dam Remnants

When the ASCE committee visited the dam in mid-June, 1889, they took along instruments and performed their own elevation survey. The elevations they reported were based on prior surveys projected along the railways over a great distance, referenced to the mean tide level at Sandy Hook, New Jersey. Francis et al. (1891) took survey measurements along the dam remnants, abutments, and on the main spillway. We used their data to construct the profile shown in Fig. 9.1, which extends

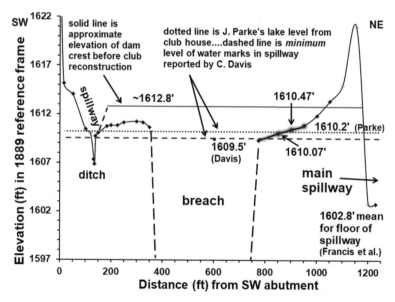

Fig. 9.1 Profile of the dam remnants and abutments from post-flood survey in 1889 (Francis et al. 1891). Data are in English units with respect to the 1889 elevation datum. The shadowed line from distance 775 ft. to 950 ft. included a 50-m length of plowed furrow that was preserved after the flood, on northeast crest (Unrau 1980 p 117; Wellington and Burt 1889). Greatest overtopping depths would have been at sagging center. A lake level of ~1610.1 ft. (1889 elevation) is estimated for the time of dam breach. Data point at the 775 ft. distance at the edge of the breach is on a slumped remnant and would not represent the elevation of the rebuilt crest. The bases for water levels by Parke, Davis, and Francis et al. are given in the text. Figure has large vertical exaggeration to show subtle differences in elevations

southwest from the main spillway along the former dam crest, across the breach, to the western dam remnant and the hillside beyond. These nineteenth century data provide the best available measurements to reconstruct the hydrologic conditions at the dam leading up to the dam breach.

Figures 4.2 (Chap. 4, riprap along dam crest) and 9.1 provide direct evidence that the dam was lowered more than 0.6 m to as much as 0.9 m by the South Fork Fishing & Hunting Club. Francis et al. (1891 p 446) claim the dam was lowered 2 ft. and report a mean height of 7.96 ft. (2.43 m) above the spillway floor for eight points on the crest of the dam remnants. I sought to recreate their mean height and got a value of 7.90 ft. using elevation stations 2, 2 + 50, 3, 3 + 50, 9, 9 + 50, 10, and 10 + 50 (see Francis et al. 1891 Plates L and LIII). However, station 10 + 50 should not have been included because it is part of the natural surface of the abutment and not the embankment. That station was 9.94 ft. above the spillway floor, near the height of the dam crest as originally built. Excluding this station yields a mean height above the spillway of 7.6 ft. More importantly, station 9 was 125 ft. (38 m) east of the breach margin and stood just 7.07 ft. above the spillway floor, as measured by Francis et al. (1891). It indicates a lowering of the dam of 0.9 m, and this part of the embankment was laterally distant from the dam center where sag due to inadequate compaction of the repaired section may have further lowered the crest. Evidence of sag is discussed in comments and diagrams by P. Brendlinger (Francis et al. 1891 p 465) who estimated that the lowest part of the embankment may only have been 6 ft. (1.8 m) above the spillway floor. Davis (1889) estimated a sag of at least 0.6 m, while others proposed even greater sag. The lowering of the crest, along with the sag, substantially reduced the originally designed spillway capacity because the relation between spillway and weir depth and discharge is nonlinear. For example, increasing the South Fork main spillway flow depth 45% (from 2 m to its original 2.9 m) would raise its discharge capacity 75%

The profile of the dam remnants has changed since the time of the flood. The two biggest changes were (1) the excavation of the northeastern breach remnant to construct a double-line railroad embankment, and (2) construction of a parking lot at the western abutment. Fig. 9.2 shows the topographic profiles of the dam remnants, comparing the 1889 topographic data with present-day GPS profiles. The top frame in Fig. 9.2 shows the superimposed profiles for the western dam remnant, along with a line that represents the approximate original dam crest. Shown in this way, it is obvious that an auxiliary spillway existed on the western abutment since even after lowering of the crest the abutment is noticeably lower than the embankment. Also see Fig. 10.3, a photograph of the present abutment that nicely shows the depth of the original auxiliary spillway. That this area of the abutment was an auxiliary spillway is further substantiated by observers noting that the initial overflow occurred here (Kaktins et al., 2013), and this is where the emergency ditch was dug. This spillway was shallower than the main spillway, but nonetheless was a critical part of the original dam's outflow capability. A hydrologic analysis of the auxiliary spillway is given in Chap. 10.

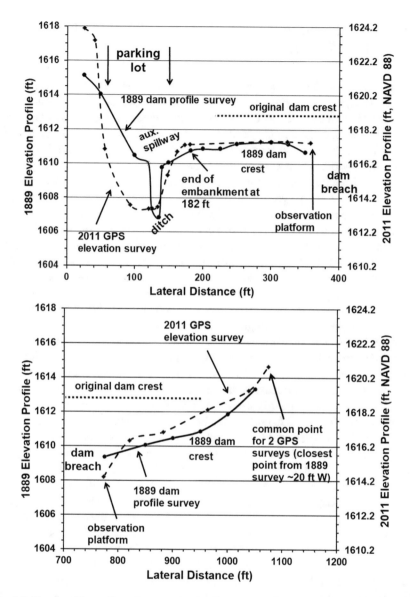

Fig. 9.2 Topographic profiles of present-day southwestern (*top*) and northeastern (*bottom*) dam remnants and abutments compared to 1889 survey. There is a difference of ~6.2 ft. (1.9 m) between the 1889 and modern elevation (NAVD 88) reference frames; the 1889 data are systematically lower (see Appendix 2). Dotted line in each frame shows approximate position of original dam crest

9.3   Water Levels for Lake and Main Spillway at Time of Dam Failure

On the day of the dam breach, workers ran a plow along the embankment crest, desperately trying to raise the lowest part of the crest and delay overtopping. This raised furrow was as much as 0.6 m high in the center of the dam and tapered to about 0.1 m at the ends. Portions of the furrow were still visible after the dam breach on the eastern remnant and provided evidence of the lake level at the time of dam failure. Wellington and Burt (1889) describe a preserved mound near the edge of the breach and extending for about 46 m on the eastern dam remnant (shadowed line segment in Fig. 9.1). The preservation of this soft mound of material, which would have rapidly been swept away by overtopping flows, shows that the lake level was no higher than 1610.1 ft. (1889 survey), and may have been somewhat less. Correcting for the difference between 1889 and modern elevations, the maximum lake level was ~1616.3 ft. (NAVD 88), or ~492.6 m. A best estimate of the lake surface elevation at the time of dam failure is in the range from 492.5 to 492.6 m.

Davis (1889) provides important information on water levels in the spillway around the time of the dam breach. When he visited the dam after the flood he could not find water levels on the bridge supports but "Out in the weir [spillway] itself I found marks on the shale banks and on the bushes, showing that the water was over 6 feet above the sill." (Unrau 1980 p 128). His uncertainty about the height of the water marks relates to his use of a hand level. The sill was a horizontal wooden beam, 6 × 6 inches, that ran along the base of the foot bridge.[2] The bridge crossed the spillway near the upstream entrance, where the surface is 0.3 m lower than at the spillway crest (175 to 190 ft. north of the bridge). Given irregularity on the spillway floor, the top of a horizontal beam would be closer to 8 inches above the lowest points. Adding the sill height to Davis' observation gives a minimum flow depth of 6.7 ft. (2.0 m) in the spillway. Adding this 6.7 ft. to Francis et al.'s (1891) mean floor elevation of 1602.8 ft. gives a water level elevation in the spillway of ~1609.5 ft. (1889 elevation, Fig. 9.1). In the present reference frame, that elevation is ~1615.7 ft. (1609.5 [Davis] + 6.2 ft), or ~492.5 m (NAVD 88). Davis' observations represent the best available information about peak flow depths in the spillway itself, which provides an estimate of critical flow depth over the spillway crest.

Another estimate of spillway flow depth was provided by John Parke, the resident engineer at the Club (Francis et al. 1891), who reported that as the breast was being broadly overtopped the water level was a little over a stake that was "...7.4 feet above the normal lake level." But that measurement was taken near the Clubhouse about one mile from the dam. It represented the lake level, not the level in the spillway. Parke did not give the elevation of "normal lake level" or state how it was measured, but this terminology usually refers to the high point on the spillway floor as a control on the lake water surface. This meant it was either "top of sill" or the spillway floor elevation, which differ by 0.2 m. If Parke's level was referenced

---

[2] The present-day bridge across the spillway has no wooden sill along its base.

to the spillway floor then there was a 0.2-m difference between his measurements and those of Davis. If Parke's measurement referred to height "above the sill," then the difference between his levels and Davis' would increase to 0.4 m. These differences are too large to be explained by the slight gradient in the lake surface between the Clubhouse and the main spillway, which would yield a height difference of less than 1 cm.

A difference between the lake level and the spillway flow level was nonetheless expected, especially when flow in the spillway was deep, caused by the physics of flow over a broad-crested weir. The lake level, i.e., upstream head above the weir crest, would have been higher than the critical depth over the crest. Even though observers like J. Parke reported there was no large amount of debris clogging the spillway mouth, additional differences between the lake level and flow level in the spillway would have been caused by flow resistance through the fish screens and around the bridge supports. The combination of rigid and floating fish screens extended at least 1.5 m up from the bottom of channel. By Brendlinger's estimate, the bridge supports and screens may have reduced the cross-sectional area available for flow by 40–50%. Old diagrams of the bridge and a post-flood photo of it in Francis et al. (1891) suggest that the reduction in flow area was not so large.

There is a simple explanation for the apparent difference between Davis' spillway marks (flow depth 6.7 ft) and Parke's inferred "...7.4 feet above the normal lake level." Parke had been working at the Club less than two months and had never seen the lake under summer-like, low- inflow conditions. Any measurements of spillway flow depth would have been taken at the foot bridge; there was no other convenient place. If he assumed that the spillway *surface* under the bridge was the high or controlling crest (normal lake level), then his water level measurements would have been overstated. The actual crest that controlled the minimum lake level was about 175 to 190 ft. along the spillway and was 0.3 m (1 foot) higher than the surface under the bridge. Parke might not have known this unless he waded in the spillway from the bridge to the crest (unlikely in that cold mountain lake in springtime). He needed to subtract ~0.7 to 1 foot from his measurement and infer a spillway flow depth of ~6.4–6.7 ft. at the time of overtopping. This is very close to Davis' observation and would reconcile most of the difference. One final point – on p. 466 of Francis et al. (1891), Brendlinger suggests that Parke's lake level measurement near the Clubhouse was relative to the "top of sill" at the bridge. However, Parke's letter to the committee does not mention the sill, only that the measure was "...a little over a stake" that was "...7.4 feet above the normal lake level."

Wind effects during the storm could have caused intermittent wash-over of the embankment near the time of overtopping. I have not found records on wind speed and direction for the dam vicinity and therefore could not incorporate wind effects in the analysis. The poorly reconstructed dam survived overtopping for >3 h, which may have included some wind-assisted wash-over. In any event, as my analyses will show, a properly rebuilt higher dam would not have been overtopped by lake rise, and hypothetical intermittent wash-overs caused by wind should have had little effect given the robust design and larger riprap.

9.4   Storage Capacity of Lake Conemaugh at the Time of Dam Failure

Coleman et al. (2016) developed a new estimate of the storage capacity of Lake Conemaugh using the 3.2 ft. DEM (digital elevation model) derived from Pennsylvania LiDAR data (PASDA 2013a, b). This DEM has spatial and vertical resolution better than 1 m. LiDAR data are scale independent and permit much more detailed mapping than the previous USGS (U.S. Geological Survey) DEMs that were based on photogrammetric analysis of decades-old imagery. At the moment the dam breached in 1889 the impoundment held about 1.455×10^7 m^3 of water below a contour elevation of 492.56 m, which approximates the modern elevation of the impounded lake at failure. The equivalent tonnage (14.3 million) is less than the usually cited figure of 20 million tons (McCullough 1968, p. 41). Previously, Penrod et al. (2006) used a DEM and GIS technology to estimate a lake volume of just over 1.27×10^7 m^3 when the dam failed. Most of the difference between that estimate and ours results from their use of a lower lake stage of 491 m for the time of dam failure (Fig. 1 of Penrod et al. 2006). A stage of 491 m is lower than the present-day, soil-covered, surface of the main spillway, based on LiDAR and GPS data. The new analysis (Coleman et al. 2016) indicates an increased lake volume of 15% at the time of dam failure and significantly adds to the time needed to drain the lake. Analysis using the LiDAR DEM also shows that the dam as originally built with a higher crest would have impounded 1.627×10^7 m^3 (16 million tons) below a lake stage of 493.5 m.

9.5   Storage–Elevation Curve for Lake Conemaugh

With LiDAR data a new storage elevation curve is developed that shows the amount of impounded water beneath any given lake level (Fig. 9.3). The shape of the curve shows that increases in stage height reflect nonlinear increases in lake volume. No correction is included for lake-bottom sediments that would have been present in 1889 and would have slightly diminished the original lake volume. There is post-flood photographic evidence that such sediments were present (Fig. 3 of Kaktins et al. 2013), but the average sediment thickness over the basin before the dam breach is not known. A significant fraction of the former lake bottom sediment probably washed away in the first years after the dam breach, before it could be stabilized by vegetation. As part of a former NPS research permit (NPS 2012), in October 2011 soil samples were collected and analyzed from five locations in the southern part of the lake basin. The samples were collected under the supervision of NPS employees. In this small sample set the soils had an average depth of 27 cm based on depth-of-refusal. Part of this soil depth would have included native soils that were present before the dam and reservoir existed.

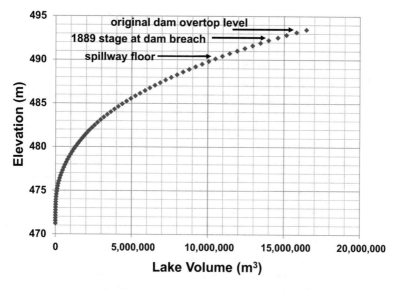

Fig. 9.3 LiDAR-based storage-elevation curve for the former Lake Conemaugh. The highest point on the curve represents an overtopping level of 493.5 m for the South Fork dam *as originally built*. LiDAR data source: PASDA (2013a, b)

9.6 Inflows to Lake Conemaugh – Time of Concentration and Time to Peak Discharge

A key factor for the survivability of the South Fork dam was the time-varying rate of flow into the reservoir. The stream flows were responding to runoff from one of the worst storms ever documented for the region, although greater precipitation depths were recorded during the 1977 Johnstown flood. Rainfall data from Pennsylvania stations are listed in Appendix 1 and mapped in Fig. 2.1 (see Chap. 2). Rainfalls in parts of the region appear to have reached or exceeded the threshold for 100-year, 24-hr storms (Fig. B-8 of USDA 1986).

Two concepts in hydrology are key to understanding how rivers and streams respond to storm and runoff events: time of concentration and time to peak discharge. The time of concentration is the time it takes for all parts of a watershed to contribute runoff to a downstream point on the river. It is usually derived from the longest tributary in that watershed, reaching to its farthest drainage divide. The time to peak discharge is the interval from the start of runoff to the peak discharge at a downstream point. In public notices of river warnings, authorities use the expression "the river will crest" at a given time at a specific place. That is the time of predicted peak river discharge *at that place*, when the river or flood level will be highest. Crest times will vary with location along the river.

The time to peak discharge (t_p) can greatly vary for different storms over the same watershed. For example, if a local downpour occurs over a small part of a big watershed, but happens near the stream measurement point, the time to peak will be very short, *much shorter* than the time of concentration (t_c). On the other hand, a localized downpour over a mountainous divide, at the upstream end of the longest tributary, will produce a time to peak very close to the time of concentration. The time to peak discharge depends on many factors, including the areal distribution of the storm (isolated or widespread), and on the duration, variability, and intensity of the precipitation. Other factors include the distribution of land slopes, the kinds and distribution of vegetation, and the infiltration capacity of the soil. Did another storm occur shortly before the big event, saturating the ground and reducing infiltration, stoking even faster runoff? Also, if snowmelt is caused by rainfall there will be a rapid runoff due to the sudden release of water from frozen storage. The bottom line here is that to truly understand the behavior of a stream or river, it is necessary to study the discharges and hydrographs produced by many storms in different seasons over a reasonably long period.

Discussion and analysis of the South Fork of the Little Conemaugh are presented in Appendix 3. Insights were also obtained from data collected downstream on the USGS gage at East Conemaugh. That gage is on the main stem of the Little Conemaugh River and monitors combined flows from the North Fork and the South Fork. Data were examined from the extreme 1977 flood. Evaluation of all the material yields a range of t_c for the South Fork basin of 3.6–7.3 h, and t_p (similar to lag time) was estimated in the range of 2.5–5.1 h. The t_p parameter highly depends on the storm characteristics. Data from the stream gage at East Conemaugh during the flood of 1977 suggest that in extreme events upstream on the South Fork t_p would be <8 h, and since t_p will often be less than t_c, it is likely that t_p would be less than 7.3 h. Therefore, if the rainfall peak on the night of May 30–31 occurred between 3 and 6 a.m., then local stream discharges would have been expected to peak and start diminishing between 11 a.m. and 2 p.m. (or earlier) on May 31st. That expectation is consistent with local stream observations around noon on that date.

9.7 Stream Level Observations

Francis et al. (1891 p 452–453) were of the opinion that the flow into the reservoir kept increasing up until the time the dam failed, "…and no doubt continued to do so for some time longer," but they offer little basis for that important claim. I found no evidence that the investigators were given the Pennsylvania Railroad (PRR 1889) testimonies, which contained observations of local streams on the day of the flood. It should also be recognized that stream gaging was in its infancy in the late 1800's. Practitioners were aware of the value of such information but simply did not have useful networks of stream gages to provide data. Historical records that did exist were mainly peak flood levels based on high water marks for various dates and places. Even the Little Conemaugh River above Johnstown had no continuous gage installed until after the disastrous 1936 flood.

Much of James Francis' experience with flooding came from his dam and canal work at Lowell, MA. He was familiar with and understood floods on the Merrimack and Concord Rivers. William Worthen also worked for an extended time at Lowell. The Merrimack typically remains near its discharge peak for several days, and elevated discharges could persist for more than a week. Fig. 9.4 is an example of a hydrograph for the Merrimack River from a gage near Lowell. This hydrograph represents the combined discharges for the Merrimack and Concord Rivers. The drainage basin above the Merrimack gage exceeds 4400 mi^2, far larger (83 times) than the ~53 mi^2 area of the South Fork basin. Hydrograph peaks recorded from larger basins will tend to be broader in time due to the longer watercourses and larger runoff totals. If James Francis and William Worthen extrapolated their experiences with the Merrimack River to the much smaller basin of the South Fork of the Little Conemaugh, it is understandable that they could have overestimated the duration of excess lake inflow on the day of the dam breach.

Francis et al. (1891) did give some consideration to the basin size above the South Fork dam and used it in calculations related to water available for runoff. A post-flood survey of the basin was performed by an engineer named Mr. Lee, who derived a watershed area of 48.6 mi^2 and a reservoir surface area of 405 acres. His map of the South Fork basin was published by *Eng. News* (1889 p 259). Lee also

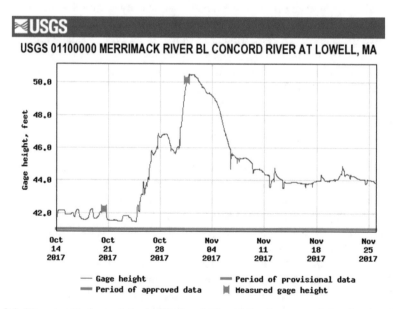

Fig. 9.4   River stage hydrograph for the Merrimack River, following a storm in late October, 2017. Gage is located 1100 ft. downstream from juncture of the Merrimack and Concord Rivers. Source: USGS (2018)

reported a useful estimate of the percentage of cleared land, which would have included areas cleared for cultivation and through timber removal:

> Judging from what could be seen of the watershed from the heights in the vicinity of the reservoir, which was perhaps half of it, it is fully half cleared land, with a gravelly soil, with little or no rock in place cropping out ... there is practically no level land in it, - all has sufficient slope to throw water rapidly off it into the streams; but taking the watershed as a whole, we should hardly say that it was likely to shed water much more rapidly than others of its size in ordinary hilly country.

Kaktins et al. (2013) document observations about local rainfall and flood levels on May 31. Much of the information comes from the statements collected by Special Agent J. H. Hampton of the PRR (1889) after the flood. The statements of eyewitnesses can be found at: http://www.nps.gov/jofl/learn/historyculture/stories.htm [Accessed 4/24/2018]. These eyewitness accounts indicate that water levels ceased to rise between 12:00 pm and 1:00 pm, and may even have started to drop in both the South Fork drainage and the main stem of the Little Conemaugh River. Flood stage in the headwaters of the main stem of the Little Conemaugh began before dawn, probably around 4:30 a.m., on May 31. About 10:00 a.m., maximum stage was reached at Lilly, ten miles from the South Fork dam. Further downstream in Wilmore, Portage, and Summerhill high stage occurred between late morning and 12:00 p.m. and the river levels began to fall in the early afternoon. At the town of South Fork, the Little Conemaugh began rising rapidly sometime before 9:00 a.m. and appears to have reached maximum flood stage between 12:00 p.m. and 1:00 p.m. Several witnesses (Allshouse and Brady) observed that the river never exceeded bank-full stage until the flood wave struck, and shortly before the dam failed the river level was observed to be "on a stand still" (Brantlinger).

Above the South Fork dam, tributaries in the South Fork drainage basin experienced high stage around 11:00 a.m. As noted in Chap. 5, testimony (PRR 1889) was provided by Station Agent C. P. Dougherty, a 25-year veteran of the Pennsylvania Railroad Co. He stated that the Little Conemaugh and the South Fork tributary had been rising steadily in the morning, but at about 10 a.m. "…they commenced to rise rapidly." He was asked whether the stream continued to rise from 1:00 p.m. onward. Dougherty replied: "I considered it falling a little. I was watching it when I had time, and I considered both streams were lowering a little which renewed a hope that it was also lowering at the dam, but the fall was hardly noticeable." These observations suggest the time of peak discharge was reached sometime between 12 noon and 1:00 p.m. His statement also revealed the time when the flood wave struck the town of South Fork. He reported that the clock at the railroad station stopped at 3:08 p.m., when the clock was thrown "out of plumb" because the floodwater floated the station off its foundation.

Downstream on the Little Conemaugh River at Conemaugh borough the water level was seen to rise until about noon and then came to a standstill, or perhaps fell slightly, in the early afternoon. Although some railroad tracks were washed out downstream, in general the river remained at about bank-full conditions in the early afternoon until the arrival of the flood wave, which indicated that runoff from the combined watersheds remained fairly constant during the early afternoon of May 31.

That timing is important because through the morning at the South Fork dam more water flowed into the lake than could flow out, causing the lake level to rise. By late morning the maximum flow capacity of the main spillway had been reached and water began to overtop the crest. Had the dam never been lowered by the Club, the pair of spillways would have provided much greater discharge capacity and the extra storage in the reservoir would have served as a protective buffer against overtopping.

There is also a record of river levels measured at Johnstown on the day of the flood. The Signal Office (Dunwoody 1889) received measurements from the Johnstown river gage on May 30th and 31st, before the gage and its unfortunate observer were swept away by the flood. At 7:44 a.m. on the morning of May 30th it read "1.0 foot above low water." At 7:44 a.m. on the 31st the reading was 14 ft.; by 10:44 a.m. the gage showed 20 ft.; and the final dramatic report at 12:14 p.m. said "Water higher than ever known; can't give exact measurement." I found no information to relate those gage readings to discharges. But the Johnstown readings show that torrential rains had occurred overnight in the Little Conemaugh watershed. The heaviest rainfall likely occurred hours before dawn because the river stage had already risen dramatically by 7:44 a.m. on the 31st.

These gage readings were reported from the Western Union Office in Johnstown, in the home of and managed by Mrs. Hettie Ogle. I have not found the exact location of the river gage, but it must have been near her office on Washington Street, close to the library. That meant that the river gage was on the Little Conemaugh River, about half a kilometer east of its intersection with the Stonycreek River. Although Dougherty had reported from South Fork around 1:00 p.m. that the "…streams were lowering a little…", the effect on the Little Conemaugh far downriver in Johnstown would have been delayed, perhaps an hour or more, but no gage records have been found later than the 12:14 p.m. report.

The stream and river observations are consistent with the analysis of the time of concentration and time to peak in the South Fork watershed. These various reports indicate that, for about 2 h before the dam failed, the flow into the reservoir was fairly constant or even decreasing. They are generally consistent with written comments by J. Parke about conditions at the reservoir (see Parke's letter to ASCE in Chap. 5). Parke later mentions that it had been raining "…most all of the morning" but never referred to heavy or intense rain during this period. Most eyewitness reports agree there was an extremely heavy rainfall overnight on May 30–31, followed by rain of varying degrees through the morning of May 31.

Based on analysis of the South Fork watershed, the time of concentration for runoff from the major rain event should have been in the range ~3.6–7.3 h, and the time to peak discharge in <7.3 h. By 6:30 a.m. the lake had already risen significantly, therefore the heaviest rainfall likely lasted for at least several pre-dawn hours, but the actual time and duration of the rainfall is not known. Presuming that the most intense rainfall occurred between 3 and 6 a.m., or earlier, the time to peak discharge for the South Fork basin should have occurred between 11 a.m. and 2:00 p.m., or earlier. The observations summarized above show that streams in the area reached their highest levels between 10:00 a.m. and 1:00 p.m., depending on location. Therefore, if the South Fork dam could have avoided overtopping for a longer period it could readily have survived. It is a significant observation that, in the actual event, the poorly reconstructed dam underwent more than *3 h of overtopping* before failure. As will be shown in the calculations of Chap. 10, had the dam been repaired to its original height, overtopping and dam breach failure would have been prevented in 1889.

References

Coleman NM, Kaktins U, Wojno S (2016) Dam-breach hydrology of the Johnstown flood of 1889 – challenging the findings of the 1891 investigation report. Heliyon 2(2016):54. https://doi.org/10.1016/j.heliyon.2016.e00120

Davis C (1889) The South Fork dam, Proceedings of Engineers' Society of Western Pennsylvania, p 89–99

Dunwoody HHC (1889) Monthly weather review, v XVII, No 4, p 117–118, Signal Office, Washington City, May 1889

Eng. News (1889) Watershed of the South Fork Reservoir, Engineering News, Sep 14 1889, p 259

Francis JB, Worthen WE, Becker MJ, Fteley A (1891). Report of the committee on the cause of the failure of the South Fork dam. *ASCE Trans* XXIV, 431–469

Kaktins U, Davis Todd C, Wojno S, Coleman NM (2013) Revisiting the timing and events leading to and causing the Johnstown flood of 1889. PA Hist J Mid-Atlantic Stud 80(3):335–363

McCullough DG (1968) The Johnstown flood. Simon & Schuster, New York, p 302

NPS (National Park Service) (2012) Investigator's annual report, permit number JOFL-2011-SCI-0001

PASDA (Pennsylvania spatial data access) (2013a) Download site for PAMAP LiDAR: http://www.pasda.psu.edu/uci/SearchResults.aspx?originator=%20&Keyword=pamap%20lidar&searchType=keyword&entry=PASDA&condition=AND&sessionID=5775552642015710143015. Accessed 24 April 2018

PASDA (Pennsylvania Spatial Data Access) (2013b) Metadata summary for PAMAP program 3.2 ft digital elev. model of Pennsylvania (based on LiDAR data). Available at: http://www.pasda.psu.edu/uci/DataSummary.aspx?dataset=1247. Accessed 4 April 2018

Penrod K, Ellsworth A, Farrell J (2006) Application of GIS to estimate the volume of the great Johnstown flood. Park Sci 24(1):7

PRR (Pennsylvania Railroad) (1889) Testimony Taken by the PRR Following the Johnstown Flood of 1889. [Statements of PRR employees and others in reference to the disaster to the passenger trains at Johnstown, taken by John H. Hampton, at his office in Pittsburgh, by request of Superintendent Robert Pitcairn; beginning July 15th, 1889.] Copy in archive of Johnstown Area Heritage Association. Many stories available online at: https://www.nps.gov/jofl/learn/historyculture/stories.htm. Accessed 4 Jan 2018

Unrau HD (1980) Historic structure report: the South Fork dam historical data, Johnstown Flood National Memorial, Pennsylvania. U.S. Department of the Interior, National Park Service, p 242, package no. 124

USDA (U.S. Dept. of Agriculture) (1986) Urban hydrology for small watersheds. Natural Res. Conservation Service, pp 164 Technical Release 55

USGS (2018) Online data portal for USGS river gage 01100000 on the Merrimack River below the Concord River at Lowell, MA. US Geological Survey, National Water Information System. https://waterdata.usgs.gov/nwis/uv?site_no=01100000. Accessed 24 April 2018

Wellington AM, Burt FB (1889) "The South Fork Dam and Johnstown Disaster," *Eng.* News and Railway Journal 21: 540–544

Chapter 10
Hydraulic Calculations

10.1 Pipe Flows

Hydraulic calculations in the form of rating curves and hydrographs are provided for the South Fork dam and Lake Conemaugh on the day of the dam breach.[1] The original dam had five large drainage pipes at its base that were used to release water during dry seasons to maintain canal boat traffic from the canal basin[2] in Johnstown to points west. Those five pipes were removed by the Club workers (Francis et al. 1891). If the pipes and culvert had been repaired to the original design they could have been used periodically to lower the lake level to make any needed repairs to the upstream face of the dam. The Club workers did not realize that another purpose of the pipes was to help offset rapid rises in lake level during floods, which would have been helpful during the storm of May 30–31, 1889. The fact that the five drainage pipes were not replaced by the Club shows that the people in charge of repairing the washed-out center section of the dam were not engineers and did not understand basic principles of reservoir maintenance.

Flow to the pipes was originally controlled from a wooden tower in the lake located near the center of the dam. There were five cast-iron pipe conduits each consisting of 12 individual segments 7 ft. long, which were joined together at bell socket joints. The effective length of each pipe segment was 6 ft. 5.5 in (Unrau 1980, Appx. K) and the total length was about 77.4 ft. The cast-iron pipes were specified to have an "in the bore" [inside] diameter of 24 in (0.61 m).

By means of Darcy's 1857 modification of the Chezy formula, Francis et al. (1891) estimated the combined discharge capacity of the 5 pipes, using a head of 70 ft. and pipe length of 100 ft. to be 543 cfs (~15 m^3 s^{-1}). It is unclear why they

---

[1] JB, Worthen WE, Becker MJ, Fteley A Report of the Committee on the cause of the failure of the South Fork dam. ASCE Trans. vol XXIV, p 431–469 (1891).

[2] The canal basin in Johnstown served as the western terminus of the Portage Railroad.

© Springer International Publishing AG, part of Springer Nature 2019
N. M. Coleman, *Johnstown's Flood of 1889*,
https://doi.org/10.1007/978-3-319-95216-1_10

chose an incorrect length of 100 ft. when in their own report, on p. 444, they quote W. Morris from his 1841 report as saying that the pipes were "...about 80 feet long."

I estimated the pipe flows using an analytical equation for the inflow capacity of a pipe spillway at full flow (Chow 1964), where

$$q = a \sqrt{\frac{2gH}{1 + K_e + K_b + K_p L}} \qquad (10.1)$$

q = discharge rate (cfs) (1 m^3 s^{-1} = 35.3 cfs)
   a = cross-sectional area of pipe (ft^2) (3.14 ft^2)
   g = surface gravity (32.2 ft. s^{-2})
   H = total head (ft)
   K_e = coefficient for entrance loss [0.5 when pipes are mounted flush and do not protrude into upstream water column]
   K_b = coefficient for bend loss (= 0 for these straight pipes)
   K_p = coefficient for pipe friction loss
   L = length of pipe (ft).

Given a cross-sectional flow area of 3.14 ft^2, a coefficient for pipe friction loss of ~0.01 (appropriate for cast iron), an entrance loss coefficient of 0.50, a total head of 70 ft. (with the dam near overtopping and the pipe intakes several feet above lake bottom), and a horizontal pipe length of 77.4 ft., each pipe could have transmitted ~4 m^3 s^{-1}. Together the five pipes would have conducted up to ~20 m^3 s^{-1} when the lake was at maximum depth. This would have been the maximum discharge rate for the dam as originally built, representing flow through unobstructed pipes. If the pipes protruded into the water column on the upstream end, the value of K_e would increase, up to ~0.78, and this would reduce the q value. However, diagrams of the control tower base, pipes, and culvert (Unrau 1980 p 48) show an upstream flush mounting (K_e = 0.5) with the downstream ends protruding into the culvert. Some flow loss may have occurred depending on the geometry of the flow control device at the upstream side of the pipes. Much of the difference between this flow calculation and that of Francis et al. (1891) was due to their use of an incorrect pipe length of 100 ft.

Debris on the screen leading to the pipe intakes could have caused additional flow losses. However, the intakes were at the lake bottom, which lessened the chances for clogging the screen. The analytical solution assumes that the downstream ends of the pipes discharged into an air-filled space, resulting in zero exit losses. Plate LI of Francis et al. (1891) confirms that this air-filled space at the upstream end of the culvert was ~4.3 m wide, narrowing to about 2.4 m. Historic construction drawings suggest the pipes were laid such that their bottoms were ~0.3–0.6 m above the culvert floor. If so, during pipe discharges the culvert volume would have been at least partly filled with water, which would have caused small exit losses and slightly reduced the discharge estimate given above. For example, if the pipe ends in the culvert were fully submerged (~0.6 m), the H value for Eq. (10.1) would be the lake depth minus 0.6 m, which would reduce the total pipe q at the time of embankment overtopping by <2%. There were no records found to ascertain the energy slope on the floor of the culvert to determine whether it could have prevented significant filling of the culvert chamber. But diagrams shown by Francis et al. (1891, plate XLVIII) suggest it was virtually level, so some submergence was likely. A phenomenon that could partly fill the culvert would be the spawning of a standing wave where the flows discharged from the pipes into the culvert, transitioning from supercritical to subcritical flow. However, as shown above, the reduction in pipe q from submergence effects would have been small.

10.2 Auxiliary Spillway

Frank (1988) presents evidence that the builders of the South Fork dam followed the original design specifications, which required the total width of the spillways be at least 46 m, and that they constructed a shallow auxiliary spillway on the western side of the dam to augment the main spillway. Frank asserted that, had the crest not been lowered and kept *as originally built*, this emergency spillway would have been more than 1.1 m deep and "…wide enough [>70 ft (21 m)] to have carried off the waters of a storm greater than the one of 1889." Frank (1988) also concluded that the dam would not have been overtopped, but unfortunately he gave no supporting discharge calculations.

If the dam's crest had been preserved by the Club at its original height, then the auxiliary spillway would indeed have functioned at high lake stages. The dimensions of this feature can no longer be measured accurately because a parking lot has been built on the western abutment to accommodate visitors. However, the 1889 topographic survey of the southwestern dam remnant and abutment area (Figs. 9.1 and 9.2) is reasonably consistent with the dimensions given by Frank (1988), although I interpret a flow depth of 0.9 m rather than 1.1 m.

The NPS has done an exemplary job of historical preservation at the South Fork dam. The museum on the hill overlooking the dam is an excellent facility to visit and learn about the history of the flood. After acquiring the property in the 1960's, one of the early protective actions was to prevent vehicles from driving through the eastern spillway (Fig. 10.1). Eventually a foot bridge was built across that spillway, similar in plan to the 1889 bridge. At the western end of that bridge, still visible in the woods after nearly 130 years, is the trace of the old road to South Fork used by the Club members. Construction of the western parking lot was done in a way that preserved key evidence of the auxiliary spillway on that abutment. Fig. 10.2 compares a photo from the 1970's with a 2018 image. The low area representing the spillway is readily seen by projecting a line along the present dam crest. This figure shows the area where the Club workers and some of their neighbors dug a ditch in 1889 to try to save the dam from overtopping. It is easy to visualize how much deeper and effective the auxiliary spillway would have been with the dam crest 0.9 m higher, as originally built in the 1850's.

Fig. 10.1 Photo from the 1970's showing protective barricade placed by the NPS to prevent vehicles from driving onto the northeast abutment and spillway. Image credit: NPS/Johnstown Flood National Memorial, Staff Photo

Flow through the emergency spillway can be estimated using the analytical equation for long, broad-crested weirs (Chow 1964, p 15–33; Dingman 1984). Analysis of these weirs is straightforward because the estimated discharges are not very sensitive to the slopes upstream or downstream from the spillway, so long as the flow depth is small compared to the weir length (d/L < 0.4) (Dingman 1984, p 230; Tracy 1957).

$$Q = CL\sqrt{g}\, d^{1.5} \tag{10.2}$$

where Q = discharge [m^3·s^{-1}]

C = dimensionless weir coefficient [~0.465 for long or normal weirs, where weir head divided by downstream length of weir crest <0.4]

g = gravitational acceleration [9.8 m·s^{-2}]

L = length of weir crest, perpendicular to flow [21 m]

d = depth of flow over weir [0.9 m]

The auxiliary spillway would have had a discharge rate (Q) of 26 m^3 s^{-1} when the lake level would have been close to overtopping the dam *as originally built*. Because the Club fatally lowered the dam crest (Fig. 10.3), this added flow capacity was not available to protect the dam on the day of the 1889 flood.

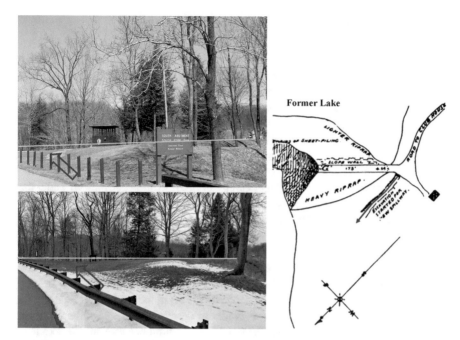

Fig. 10.2 Photos of western abutment, South Fork dam, with 1889 sketch. *Top left*: circa 1970's (image credit: NPS/Johnstown Flood National Memorial, Staff Photo). *Bottom left*: Photo March 2018 by the author. Yellow lines are extensions of the crest of the western dam remnant. Low area at left (in both photos) represents part of auxiliary spillway. Blue arrows show flow direction at high lake level. Sketch at right from surveys by *Eng. News* staff (1889 p 541). Their plot of excavation (ditch) on abutment is consistent with low area seen today

Fig. 10.3 View of southwestern abutment in 2017. Bottom blue line is extension of the dam crest as seen today, similar to the crest in 1889. Upper blue line represents height of crest before it was lowered by the Club, showing the width and depth of the auxiliary spillway excavated here in the original dam construction. Photo by the author

10.3   Main Spillway and Combined Discharge

Equation (10.2) was used to analyze the flow capacity of the main spillway for two cases: flow depth and discharge for the reconstructed dam, and a hypothetical evaluation of the same parameters for the dam at its original height. In both cases a flow width of 21 m as measured 1 m above the spillway floor was used, based on historical information (Francis et al. 1891, p 455). My own site measurement with Professor Kaktins showed the present width at the narrow portion is ~1.5 m wider at a height of 1 m above the spillway floor, probably due to 125 year. of erosion of the western margin of the spillway. The eastern margin is a robust highwall that exposes sandstone beds of the Conemaugh Group. Elevations in Fig. 9.1 show that the main spillway for the reconstructed dam had a flow depth of ~2.0 m (1609.5 ft. from Davis [1889] minus 1602.8 ft.; spillway floor). Using eq. (10.2) a discharge of ~86 m^3 s^{-1} is obtained for the main spillway of the reconstructed dam, given $d = 2$ m and $L = 21$ m.

Francis et al. (1891) used three methods to estimate the discharge in the main spillway. Their preferred method gave a discharge rate of 3700 cfs (~105 m^3 s^{-1}), but they used an average flow depth of 7.5 ft. (2.3 m) in the spillway (Francis et al., 1891, p. 452). This flow depth corresponds to a water surface in the spillway at 1610.3 ft. (1602.8 + 7.5) (see Fig. 9.1), but it is too high because such a lake level would have washed away most of the plowed furrow on the crest of the eastern dam remnant. But much of the furrow remained, and this would have been visible when Francis and his colleagues visited the site. Their flow depth is also nearly a foot higher than the spillway water marks reported by Davis (1889). Using a more realistic flow depth of 2.0 m, their three methods would yield a range of discharges from 74 to 136 m^3 s^{-1}.

Two things could have caused the discharge rates to be somewhat less than the estimates noted above. First, Plate XLVI of the ASCE report shows that the bed of the spillway was very rough after the flood, and this would have caused energy losses. Any roughness of the original spillway floor would have been enhanced by erosion during the flood flows. Some additional energy loss would have resulted from the main spillway being curved. The discharge rates in both scenarios (original vs. reconstructed dam) would be affected by these factors, and thus no explicit corrections were needed to make comparisons. Other factors such as the effects of flow convergence at the entries to both the main and auxiliary spillways would have been minimal due to the curved margins. Also, the spillways would not have been subject to submergence effects at their downstream ends because both spillways have long crests and terminated at steep slopes.

The spillway was much wider at its entry near the foot bridge than at its narrowest part, about 200 ft. beyond the bridge. This led Francis et al. (1891) and John Parke to conclude that the obstructions at the entry to the main spillway (i.e., fish screens and bridge supports) did not significantly impede the outflow, based on the assumption that the narrowest part of the spillway controlled the discharge rate. For comparison with analysis by Francis et al., I also evaluated the spillway flow without consideration of these obstructions.

Table 10.1 Added discharge capacity of the dam as originally built

Auxiliary spillway	~ 26 m^3 s^{-1}
5 discharge pipes	~ 20 m^3 s^{-1}
Main spillway	~ 65 m^3 s^{-1} (i.e., 151 minus 86 m^3 s^{-1})
Total *added* discharge	**~ 111 m^3 s^{-1} (3918 cfs)**

Prior to overtopping, the discharge from the Club's reconstructed dam occurred mostly through the main spillway. A small discharge passed through the newly dug ditch at the southwestern end of the dam. In an interview, Parke noted there was about three feet of rock through which it was possible to cut before striking bedrock (*Pittsburgh Commercial Gazette* 1889). However, in his formal letter to the committee (Francis et al. 1891, p 449–450), Parke stated that the ditch was dug 2 ft. wide and 14 inches deep into original ground (abutment) about 25 ft. from the constructed part of the dam breast. He estimated an eventual flow depth of ~20 inches in the deepest part of the ditch. More than three feet of excavation would be consistent with the 1889 survey data of Francis et al. (1891). But P. F. Brendlinger[3] did not see evidence of this ditch when he visited the site a year after the dam breach (Francis et al. 1891, p 464).

Had the dam been rebuilt to its original height the maximum flow depth in the main spillway at overtopping would have been ~3 m. This is the 10 ft. freeboard for spillways from Morris' dam specifications. Compared to the overtopping lake level, I use a lower flow depth in the spillway due to critical flow over the spillway crest, ~2.9 m. Applying eq. 10.2 to estimate main spillway discharge gives a result of ~151 m^3 s^{-1}. This discharge rate is 76% greater than my estimate for the main spillway of the Club's reconstructed dam. The *added* discharge capacity of the dam had it been kept at its original design is shown in Table 10.1. The Club workers and officers never understood that the changes they made eliminated the auxiliary spillway and cut the safe discharge capacity of the dam in half. That is the tragic consequence of not hiring a qualified engineer to consult on the project. The dam was now doomed.

10.4   Inflow to Lake Conemaugh

The only actual measurement concerning the rate of rise just prior to overtopping was provided by J. Parke. In a newspaper interview several days after the flood he stated that the lake water was rising "...at the rate of about ten inches an hour"

---

[3] Brendlinger attended the Cresson convention in 1890 and while there traveled on his own to the South Fork dam where he closely examined the dam remnants. The organizing committee for the meeting included Club member R. Pitcairn. No excursion to the dam was included in the meeting itinerary. Brendlinger reported his observations about the dam when the South Fork dam report was presented by Francis at the Chattanooga meeting in 1891.

(*Pittsburgh Commercial Gazette* 1889). In the same article he noted when he awoke that morning (~6:30 a.m.; Francis et al. 1891 p 448) he found the lake had risen "... until it was only four feet below the top of the dam." That suggests an *average* lake rise over 5 h of ~9½ inches/hr. from early morning until the dam overtopped. Parke gave a late morning rate of rise in the formal letter he wrote to the ASCE committee on August 22, 1889 (Francis et al. 1891, p 448–451). He wrote that "...the lake in the hour had risen 9 inches." That observation was likely made sometime between 11:00 and 11:30 a.m., just before the water began overtopping the dam. The reliability of this observation is uncertain, but it is consistent with the rate of rise through the morning prior to overtopping. Parke's estimate is the only available information that can be used to estimate the inflow rate to the lake at a specific time. No gaging data existed for the streams above Lake Conemaugh. The fact that Parke's letter gave the value 9 inches rather than a rounded figure like "about a foot" suggests that some care was used to obtain the estimate. Parke's estimates should be given credence because the work he had been doing for the Club required careful measurements of topographic slope while he was supervising the digging of drainage channels.

Note that the rate of rise would have been less than 9 inches per hour if five large drainage pipes had been kept beneath the dam when the Club repaired it. Those pipes would have been available for added discharge during floods. I address uncertainty in Parke's lake rise estimate by including an additional 15 m^3 s^{-1} in the estimate of excess inflow to the lake. Also, lake rise calculations presented here for the dam *as originally built* make the extreme assumption that high inflows continued unabated through the day, even though streams had reportedly ceased to rise.

Using the storage-elevation curve for the lake basin (see Chap. 9, Fig. 9.3) I calculate how much excess inflow to the lake, i.e. beyond spillway capacity, would have been needed to produce the 9-inch (23-cm) rate of rise. At the time of the observation the rebuilt dam was close to overtopping and the lake surface elevation would have been less than 492.5 m (NAVD 88). At this lake stage a 9-inch (23-cm) rise equates to an added lake volume of ~4.14 × 10^5 m^3 or ~115 m^3 s^{-1} as excess inflow to the lake beyond the capacity of the main spillway. According to Parke's narrative most of the 9-inch rise occurred before flow began in the emergency ditch (Francis et al. 1891, p 449). The excess inflow of ~115 m^3 s^{-1} is only slightly more than the *added* capacity of 111 m^3 s^{-1} that would have been available for the dam as originally built (Table 10.1). Therefore, if Parke's estimate was approximately correct and close to the time of peak inflow, the design of the original dam would certainly have averted overtopping long enough for stream inflows to diminish. My calculations of lake rise assume some uncertainty in Parke's estimate and therefore use the larger value of ~130 m^3 s^{-1} as excess inflow to the lake. The calculations will show how overtopping would have been significantly delayed by the greater discharge capacity that would have been available for the original, higher dam crest. Therefore, even if the inflow inferred from Parke's rate of rise had been maintained for *many* hours, the lake would not have overtopped the embankment as originally built. As previously indicated, there is evidence from local observers that in fact the streams had ceased to rise. Parke himself states that at about noon the lake level was

"...almost at a stand..." due to the added outflow from overtopping. That was 3 h before the dam breach. And in an interview given just 4 days after the flood, Parke stated that when he walked over the dam at about 1 p.m. (1.5 h after overtopping began), there were only about 3 inches (8 cm) of water going over the dam (*Pittsburgh Commercial Gazette* 1889).

It was challenging to determine the depth of water going over the crest of the dam because most observers, including Parke, did not specify where along the embankment that depth was observed, although one can presume it was the dam center, nor do they always specify the time of observation. Most estimates of water depth by observers are for a foot or less. For example, the ASCE report (Francis et al. 1891 p 454) estimates that water depth over the dam was 1 ft. deep for a length of 100 ft. (30.5 m), and about 9 inches deep for a length of 300 ft. just prior to the break. Parke does state that the water was overflowing about 300 ft. of the crest, but this was around 11:30 a.m., more than 3 h before the breach. The scenario of about 400 ft. of overflow just prior to failure also seems improbable because it implies the dam was broadly cut down from the top by water flowing over it (the breach width was ~420 ft) and there is credible evidence that this was not the case. There is no evidence in Parke's statements of broad, substantial downcutting of the embankment crest. He instead observed that a narrow "hole" about 10 ft. wide and 4 ft. deep was eventually cut into the outer face. Washovers through this trough severely eroded the downstream face and toe of the dam, laterally thinning the embankment until "the pressure of the water broke through" (Francis et al. 1891 p 451) creating a rapidly growing trough in the dam breast. Flow through this breach was so great that Parke reported he saw a $\geq$ 10 ft. drawdown in the lake surface that extended about 150 ft. back from the crest.

The highest overtopping estimate of 16 inches was given by A. Y. Lee (Unrau 1980 p 134) as the depth when the dam failed, but it is not clear how he determined this. However, if Schwartzentruver's statement (Russell 1964) many years after the flood is indeed accurate, that he and his young friends crossed the dam on foot shortly before failure, then it is highly improbable that water depth on the crest could have been 16 in, or even 1 ft. on the central portion. Parke's "real time" estimate of perhaps 3 inches is much more likely, prior to the formation of the 10 foot trough. At the very least this supports Parke's conclusion that the water level was nearly at a stand. Therefore it appears that the stream inflow had peaked around noon, which would be consistent with the assessment of the time to peak discharge for the South Fork of the Little Conemaugh River (see Appendix 3). Accordingly, the extreme rate of rise reported by Parke for late morning was not sustained and probably represents the general time of peak inflow to the reservoir.

Fig. 10.4 shows the discharge capacity of the dam via its two spillways and five sluice pipes, representing the dam as it was originally built. Assuming the dam had been rebuilt to its original specifications. Two vertical black lines show the overtopping lake levels for the original dam and the dam as rebuilt and lowered by the Club. Table 10.2 summarizes key data for the dam and watershed. Analysis shows that the lake surface at time of dam failure had an elevation of ~492.5 to 492.6 m (NAVD 88). The dam as originally built could have discharged floodwater at a maximum

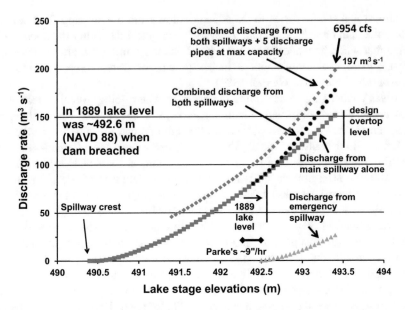

Fig. 10.4 Discharge capacity for the South Fork dam as originally built, with embankment ~0.9 m higher. The base level of this chart is ~490.4 m, which is the modern-day elevation of the subsoil spillway crest (see Appendix 2 and Coleman et al. 2016). Overtopping would have occurred at ~493.5 m (NAVD 88) if the dam had been rebuilt to its original height. The spillway flow level would have been ~0.1 m less

rate of ~197 m³ s⁻¹, given a spillway flow surface elevation of 493.4 m (NAVD 88), more than twice greater than the estimated discharge capacity of the reconstructed dam. This greater discharge capacity has big implications for the survivability of the dam *as originally built*.

10.5   Inflow Calculations by the Investigation Committee (Francis et al., 1891)

Part of this retrospective analysis of the ASCE committee's conclusions is an evaluation of their lake inflow calculations, which led them to believe the dam would have been overtopped and destroyed even had it been rebuilt as originally constructed. Francis et al. (1891 p 452–453) give three estimates of the rate of rise of the lake stage and present the calculations as evidence that the rate of lake inflow continued to increase until the dam failed. These are shown in Fig. 10.5. Their first estimate was based on a rise in lake level of 10 inches/hr. around 10 a.m. They assumed a spillway flow depth of 5 ft. and calculated the total inflow to the lake at 7208 cfs (204 m³ s⁻¹). No reference was cited for the source of the 10 inch rise per hour or for the reported 10 a.m. timing. In Col. Unger's later testimony for the Pennsylvania Railroad, he stated that the lake was rising "at the rate of about ten

Table 10.2 Summary of key data for the South Fork dam and watershed

Dam comparisons	Original dam	Rebuilt dam
Embankment height	72 ft. (22 m)	<70 ft. (~68–69 ft. in center)
Overtopping elevation (NAVD 88)	~493.5 m	~492.6 m (~1616 ft) (lake level at time of dam breach)
Discharge capacity (Q) at overtopping		
Main spillway	~151 m³ s⁻¹	~86 m³ s⁻¹
Auxiliary spillway	~26 m³ s⁻¹	Ditch ~ <10 m³ s⁻¹
Five discharge pipes	~20 m³ s⁻¹	0
Total Q	~197 m³ s⁻¹ (6954 cfs)	~96 m³ s⁻¹ (3430 cfs)
Estimated time of overtopping	Not overtopped	~11:30 a.m. (actual)
Hydraulic and basin data		
Basin drainage area above lake	~53 mi² (137 km²) (Brua 1978)	
Spillway floor elevation (NAVD 88)	~490.4 m (crest)	
Time of concentration, south fork of the little Conemaugh R.	3.6 to 7.3 h	
Time to peak inflow to lake after period of most intense rainfall	2.5 to 5.1 h (≤7.3 h in extreme events)	
Rate of lake rise ~10:30–11:30 a.m. just before initial overtopping of dam	~9 in. hr.⁻¹ (J. Parke; Francis et al. 1891 p 449) [corresponds to inflow of 115 m³ s⁻¹ beyond main spillway capacity of rebuilt dam]	
Estimated peak lake inflow before 11:30 a.m.	201 m³ s⁻¹ [spillway (86) + excess inflow (115)] [inflow of 216 m³ s⁻¹ used in this analysis]	
Lake volume at time of 1889 dam breach	1.455 × 10⁷ m³ (below stage 492.56 m, NAVD88)	
Time to drain lake to base of the upper breach	≥ 65 min	

inches an hour" at about 6 a.m. At 6:30 a.m. John Parke (Francis et al. 1891, p 448) noticed that the lake had risen about 2 feet during the night, went to get breakfast and after eating returned to the lake and found that it had risen 4 or 5 inches during breakfast. Perhaps the committee estimated ½ hr. for Parke's breakfast and ~10 in hr.⁻¹ rise, or they saw the newspaper account of his early morning estimate of about 10 in. hr.⁻¹ rise (*Pittsburgh Commercial Gazette* 1889). However, these lake rise estimates were much earlier than 10:00 a.m. Parke reported a rise of 9 inches in an hour just before the lake began overtopping the dam.

Francis et al. (1891 p 452) based another total inflow estimate of 7980 cfs (226 m³ s⁻¹) on Parke's reported 9 inch per hour rise from about 10:30 to 11:30 a.m., based on a spillway flow depth of 7.5 ft. which, as previously discussed, is improbable. A spillway depth of 6.7 ft. is more consistent with observations and yields a lower inflow rate. Francis et al. (1891) also evaluated a scenario around 10 a.m. with

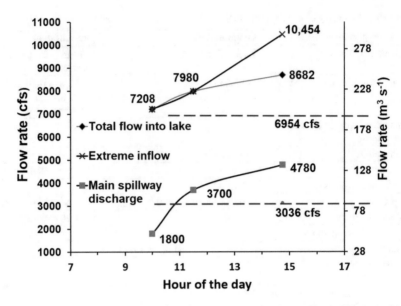

Fig. 10.5 Various inflow rates estimated by Francis et al. (1891) for the South Fork dam and reservoir. Upper two plots represent total estimated inflows to reservoir. Lower plot gives estimated discharges through main spillway. Lower dashed blue line is maximum discharge in main spillway based on 6.7 ft. (2.0 m) height of flood marks measured by Davis (1889). I estimate 86 m³ s⁻¹ for the main spillway at that time. Upper dashed line shows the total discharge capacity of the dam *as originally built* (197 m³ s⁻¹)

a spillway depth of 5 ft., and with the lake level reportedly rising at 10 in hr.⁻¹. But at this rate, in the 90 min from 10 a.m. until 11:30 a.m., the water level could not have risen 2.5 ft. (7.5 minus 5 ft). It would only have risen ~15 in, and probably less after taking into account the nonlinearity of the storage-elevation curve (Fig. 9.3).

I also examined the third and most critical calculation of inflow to the reservoir by Francis et al. (1891). They estimated a water rise of 6 in per hour *in the hour before the dam breach* (i.e., the hour before 2:55 p.m.), implying an excess lake inflow of about 2911 cfs. The dam crest was being actively overtopped at that time. They calculated that the main spillway was discharging 4780 cfs using a flow depth of 8.7 ft. Overtopping flow over the dam crest was estimated at 991 cfs, resulting in a combined lake inflow rate of 8682 cfs (247 m³ s⁻¹). As further support, the committee cited Parke (Francis et al., 1891, p. 451) who stated that after the lake drained there remained "…a violent mountain stream 4 or 5 feet deep…" that persisted until the next day.

However, the Committee's spillway flow depth of 8.7 ft. was implausible. The assumed flow depth of 8.7 ft. in the spillway is much greater than the observed 6.7 ft. (Davis) and would have required unrealistically deep overtopping levels at the dam crest that would have obliterated all evidence of the raised furrow on the eastern dam remnant. That furrow was preserved, and therefore the extreme third inflow estimate by the committee cannot be correct. The entire western dam remnant would also have been submerged. If the spillway had been flowing with a depth of 8.7 ft., the lake stage would have risen to ≥1611.5 ft. (1889 reference frame) and the entire dam would have been overtopped to an average depth of >1.5 ft. based on the survey measurements the committee themselves collected (Fig. 9.1). Even the highest points on the western dam remnant would have been overtopped by ~0.3 ft. If we consider only the top width of the breach (420 ft), hydraulic calculations show that a mean flow depth of >1.5 ft. would yield an unrealistic discharge of >1760 cfs (50 m^3 s^{-1}) just for the overtopping flow, not the 990 cfs estimated by the committee. Clearly the entire dam was never overtopped, and there is no evidence that the lake stage rose half a foot (0.15 m) in the final hour before the dam breached, that is, after ~1:50 p.m. As noted earlier, Parke reported that after overtopping the lake was virtually at a stand due to the added discharge over the dam crest.

Interestingly, the committee did a further calculation (Fig. 10.5) to estimate an upper limit for the inflow that day. They assumed a rainfall rate of 2/3 inch (1.7 cm) per hour on the watershed (20,900 cfs) and that half of this influx (10,450 cfs) would discharge into the lake (Francis et al. 1891 p 454). The committee predicted that a flow rate of this magnitude or higher could have entered the reservoir by 4 p.m. or later on May 31, more than an hour after the actual breach. This calculation seems intended to show that water levels could only have been expected to go higher. However, the committee did not consider what such an influx would mean with respect to spillway flow depths and crest overtopping levels, which likewise render this extreme influx scenario implausible.

The preponderance of evidence eliminates these extreme influx scenarios: (1) upper flood stage marks in the main spillway (Davis 1889), (2) preservation of furrow mounds on the eastern dam remnant (Wellington and Burt 1889), (3) observations that the stream levels had peaked or even begun to drop, and (4) the assessment of a plausible time of concentration and time to peak discharge for the watershed that is consistent with the stream level observations. The fact that both the South Fork and Little Conemaugh had ceased to rise or had begun to recede about 2 h before the dam breach apparently not known by the committee.

And finally, based on breach flow calculations in the next section, the committee's extreme inflow estimate (~300 m^3 s^{-1}) would have sustained a flow depth in the inner breach of ~5.7 m (19 ft) after the lake had drained. This flow depth far exceeds the observation by Parke (Francis et al. 1891 p 451) that it was "…a stream 4 or 5 *feet* deep…" Therefore the committee's extreme inflow estimate is inconsistent with present calculations and Parke's post-flood stream observation.

10.6   Dam Breach Calculations

Calculations in this section show estimates for the maximum flow rate of water through the dam breach. Hydrographs are presented to show how the flow rate and lake level diminished during the progress of the flood. The commonly reported time of 45 min to drain the lake is incorrect. The lake took more than an hour to drain.

10.6.1   Peak Discharge Rate

The center of the South Fork dam failed catastrophically more than 3 h after over-topping began. Dam breach methods have been used by other authors to estimate the peak discharge. The scientific literature contains several estimates of peak discharge rate for the Johnstown Flood of 1889. MacDonald and Langridge-Monopolis (1984) give a range of 5600 to 8500 m^3 s^{-1}, but their source is Pagan (1974), who gives the same range but no reference or method of calculation. Froehlich (1995) presents estimated peak outflows from a series of breached embankment dams, including the South Fork dam. He also gives a peak discharge rate of 8500 m^3 s^{-1} and indicates the estimate is based on reservoir volume change over a 30-min period. Froehlich (1995) identifies the rebuilt South Fork dam as having "homogeneous earthfill," which is partly correct. As designed, the embankment was a zoned earthfill dam with puddle layers on the upstream half. The repaired center section consisted of randomly placed fill and was not zoned in the same way. Froehlich (1995) reported a lake volume of 1.89×10^7 m^3 and a height of water above the breach bottom of 24.6 m. This latter number is not correct as the depth was ~21.3 m, and our LiDAR-based volume is 1.455×10^7 m^3. Singh (1996) reports a lake volume of 1.90×10^7 m^3 and flow duration of 45 min. In modeled hydrographs Singh (1996 Fig. 6.4) shows peak discharges >7000 m^3 s^{-1}. Pierce et al. (2010) give a peak outflow of 8500 m^3 s^{-1}, citing Wahl (1998), but they also use his excessively large water depth of 24.6 m.

The breach is the South Fork dam is shown in Fig. 10.6. The maximum possible discharge rate for the flood would have occurred if the entire breach formed suddenly with the lake at maximum overflow depth. To evaluate the peak discharges in the literature, Eq. 11b of Walder and O'Connor (1997) was applied to estimate peak flow through the breach,

$$Q_o = \left[c_1 r \left(D_c - b_f \right) + c_2 h \cdot cot\theta \right] g^{1/2} h^{3/2} \qquad (10.3)$$

Where Q_o = discharge from lake (m^3 s^{-1})
r = bottom width of breach (m) divided by breach depth (m) (dimensionless)
D_c = height of dam relative to base (m)
b_f = height of breach floor relative to dam base at end of flood
θ = slope of breach sides (deg.)
g = surface gravity (9.81 m/s^2)
h = lake level relative to breach floor (m)
c_1, c_2 = numerical constants related to breach shape (c_1 = 0.405 and c_2 = 0.544); represent case that neglects energy loss as floodwater approaches the breach, and also assume no tail-water effects at the outlet (Walder and O'Connor 1997).

Equation 10.3 was applied to estimate the peak discharge, treating both the "inner" breach (center of the breach; see Fig. 10.6) and "outer" breach (breach segments outside the 17 m wide center) as separate calculations. As surveyed in 1889, the top, or maximum, total width of the breach was 129.6 m. The upper part of the

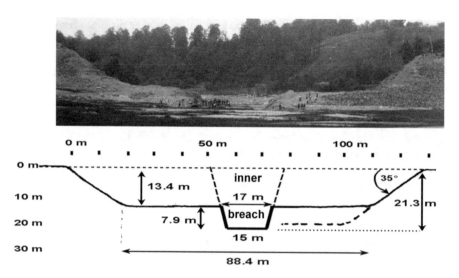

Fig. 10.6 Geometry of the breach in the South Fork dam. There is no vertical exaggeration. Dimensions based on post-flood dam survey data (Francis et al. 1891) and 1889 image (top). Dashed line at lower right shows where dam remnant was later removed to build a double rail line through the old breach

breach had an average depth of ~13.4 m and a bottom width of ~88.4 m. Subtracting the mean width of the inner breach gives a width of 88.4–19 m = 69.4 m ($r = 5.18$ and $\theta = 35°$). The inner breach was cut to the base of the dam and was ~15 m wide at the base and more than 21 m high ($r = 0.70$ and $\theta = 73°$). The combined dam breach discharge of ~8870 m³·s⁻¹ approximates the maximum possible discharge that could have occurred through the breach. At the moment of general dam failure, flow was also occurring through the main spillway and ditch, plus flow over the central portions of the crest. There is no explicit addition of overtopping flow because that is incorporated in the breach flood discharge. The main spillway transmitted ~86 m³·s⁻¹ and the ditch about 10 m³ s⁻¹, resulting in a combined flood discharge of ~8970 m³·s⁻¹. To incorporate uncertainty I include a 15% increase in breach discharge rates. Sources of uncertainty may include complexity of flow through the unusual "double" breach, accuracy of breach reconstruction from nineteenth century data and images, and the possibility that a transient deeper channel formed in the lower breach that was infilled during the waning flood. The latter point is less likely because the original dam base was excavated to bedrock. The result is a maximum peak discharge of 10,200 m³ s⁻¹ through the breach, plus ~100 m³ s⁻¹ flow through the spillway and ditch. The lake level would have rapidly fallen below the floors of the spillway and ditch.

There is little doubt that the upper part of the dam breach formed very rapidly (Kaktins et al. 2013). This may not have been the case for the bottom 7.9 m of the inner breach as its erosion through the original puddle layers at the base of the dam may have been more gradual. In that case the instantaneous peak produced by the initial flood wave would incorporate rapid failure of the upper 13.4 m of the breach (7100 m^3·s^{-1}) and flows through the main spillway (86 m^3·s^{-1}) and ditch. The result would be a peak discharge (Q_p) of ~7200 m^3·s^{-1} for the instantaneous discharge rate during catastrophic formation of the upper breach.

To summarize, the peak discharge rate (Q_p) is estimated in the range from ~7200 m^3·s^{-1} to ~8970 m^3·s^{-1}. This represents maximum flow at or just after the general dam breach and includes flow through the breach plus flows through the main spillway and ditch at that early time. Adding a 15% increase in breach discharge rates to incorporate uncertainty would yield a maximum peak discharge of 10,300 m^3 s^{-1} through the breach, spillway, and ditch.

10.6.2   Hydrographs of the 1889 Dam Breach

The mathematical model of Walder & O'Connor (1997) was also used to generate discharge and lake stage hydrographs of the dam breach flood. As shown in Fig. 10.6, the dam breach can be viewed as two trapezoids comprising an inner breach and two "wings" of an outer breach. Only the smaller inner breach eroded to the base of the dam. This complex breach prevents the use of most existing analytical models that assume simple geometric shapes. To generate hydrographs I modeled the two trapezoids as independent breaches that simultaneously drained the lake, along with residual flow through the spillway during the early minutes. The breach geometry was greatly simplified at mid to late times when all flow was in the bottom of the inner breach with an initial depth of 7.9 m. To analyze this portion of the hydrographs I applied equations from Walder and O'Connor (1997). These included their lake stage differential eq. 10b (not reproduced here) and weir eq. 11b (Eq. 10.3 above).

Two principal scenarios were analyzed. In the first the entire complex breach forms fast, such that the breach fully forms before significant drawdown of the lake. This produces the highest peak discharge and drains the lake in the shortest time. In the second scenario the outer breach and upper part of the inner breach form instantly. This condition establishes the lower range of peak discharge. The zone between these scenarios encompasses all hydrographs that would result from gradual erosion of the lower 7.9 m of the inner breach.

That the upper part of the complex breach did form almost instantly is supported by the eyewitness account of U. Ed Schwartzentruver, retold many years later in an interview, that trees immediately below the dam were felled by an air blast before the flood wave reached them (Russell 1964; Kaktins et al. 2013). Uprooting of hardwood trees by wind would require wind speeds >75 mph (34 m s^{-1}) (NOAA 2012), which was faster than the flood wave. Lesser winds may have sufficed given the deep saturation of soils from heavy rains. Wind speeds would also have been enhanced near the leading edge of a dam breach flood wave. Some people along the flood path reported high winds as the flood wave approached, including the intriguing report by Freight Conductor Fred Brantlinger. He was on one of the trains delayed in South Fork. In the PRR testimony (1889) he was asked, "Did the flood strike you when you were under way?" Brantlinger replied,

> When I jumped off the engine, it wasn't quite at me yet, and I went for the hill, and it was so slippery when I would get up a piece I would slip back; the second time the water got to me, and of course I went back the second time, and then a big wave came right in there, and it raised me right up and I got hold of a limb of a tree, and as the water raised, I went along from one brush to another. It was horrible, I tell you. There was a draught of very strong air ahead of the water, that I believe was most worse than the water, for when that air struck me, it seemed to lift me right off of the ground. A man could see it; it just looked like a blue heat you know. I don't know whether it was imagination or not but I thought I could see things falling before the water got to them. It made a terrible noise; you would think the whole earth was being torn up. The rest of the fellows had all been away from the train, and they got away nice to the hill.

Flood hydrographs were developed by summing the discharges from the inner and outer breaches over time as the lake gradually drains (Fig. 10.7). The model fully incorporates the complex breach shape in the South Fork dam and is based on the storage-elevation curve of the reservoir (Fig. 9.3) and a dam breach weir equation that allows specification of the breach side slopes. Eq. (10.3) was used with data from the storage-elevation curve by numerically evaluating lake stages and volumes that correspond with integrated model discharges through the breaches. One scenario includes a 15% increase in initial breach discharge rate to incorporate uncertainty in the overall analysis.

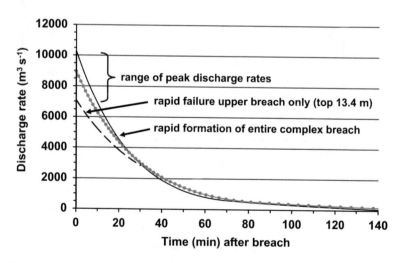

Fig. 10.7 Dam breach discharge hydrographs for the 1889 flood. Red dotted line represents scenario where the entire breach formed almost instantly. Solid black line includes 15% increase in initial discharge to incorporate uncertainty (see text). These calculations represent maximum theoretical discharge through the breach and spillway, with Q_p ~8970 to 10,300 m^3 s^{-1}. Dashed black line represents early-time discharge through only the top 13.4 m of the breach. This line intersects the vertical axis at ~7100 m^3 s^{-1}. Adding ~100 m^3 s^{-1} for the spillway and ditch yields the minimum Q_p of ~7200 m^3 s^{-1}

A hydrograph plot was included to represent early-time (0 to 30 min) discharge through the top 13.4 m of the breach, assuming it formed instantly (dashed black line in Fig. 10.7). The lowest part of the "inner" breach may have formed gradually. Hydrographs for all scenarios of erosion (rapid or gradual) of the lower 7.9 m of the inner breach would lie between the red line and the dashed black line in Fig. 10.7.

It is emphasized that the discharge hydrographs represent conditions at the breach and for the stream immediately below the dam and upstream from the juncture of the South Fork with the Little Conemaugh River. The present work does not analyze the flood routing along the Little Conemaugh to Johnstown.

Lake stage hydrographs are shown in Fig. 10.8 and illustrate the decline in lake level during the flood. Plots for three lake inflow rates are shown. Spillway flow is incorporated only in the early minutes before the lake fell below its crest. The plot for total inflow of 216 m³ s⁻¹ corresponds to the lake rise reported by Parke (9 in hr.⁻¹) just before overtopping, plus an added inflow of 15 m³ s⁻¹ to address uncertainty. Curves for two lesser inflows are shown because the lake inflow likely declined during the more than 3 h from the time of initial overtopping until the dam failed. At least 65 min were needed for the stage to fall below the floor of the upper breach. At late times the lake levels approach equilibrium as discharges through the lower breach approach the lake inflow rates. To help verify the calculations for the

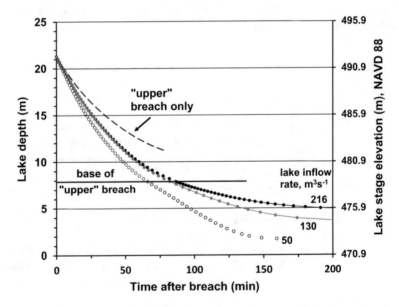

Fig. 10.8 Lake stage hydrographs for Johnstown Flood of 1889. Curves are plotted for three different lake inflow rates. Dotted lines represent cases where the entire complex breach formed rapidly, which would have drained the lake in the shortest time. Plot of open circles includes 15% increase in initial discharge rate to incorporate uncertainty. Dashed line shows hypothetical lake stages if only the upper 13.4 m of the breach had formed. The zone between the dashed and dotted lines represents a family of curves for all plausible erosion times for the lower 7.9 m of the inner breach

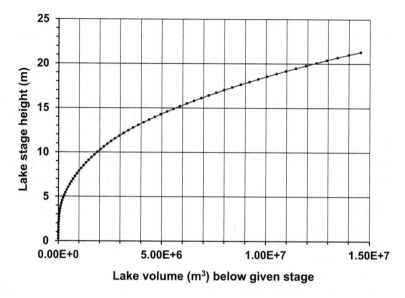

Fig. 10.9 Calibration plot shows good water-balance fit between the storage-elevation curve for Lake Conemaugh (black data points) and total lake volume minus cumulative model discharge for the rapid breach scenario (curved black line). Model discharges used to prepare this plot do not include lake inflow. The curve match is consistent with a lake floor shape factor (*m*) of 2.7, representing a basin with gently sloping sides and bottom (see Walder & O'Connor 1997 p 2341)

discharge and stage hydrographs, Fig. 10.9 was prepared to demonstrate the close water balance in the model. This calibration plot compares cumulative model flood discharge with the storage-elevation curve for Lake Conemaugh.

The time of dam breach (time zero for the hydrographs) occurred between 2:50 and 2:55 p.m. (Kaktins et al. 2013). It took 13–18 min for the initial flood wave to reach the rail station at South Fork, where the flood knocked the clock off plumb at 3:08. McCullough (1968 p 147) wrote that the generally accepted time the flood wave reached Johnstown was 4:07 p.m. These times permit calculation of the average speed of the flood from the South Fork Dam until it struck Johnstown. The river path from the breached dam to the outskirts of Johnstown is 23.5 km long, which includes the distances around the river loops at the Conemaugh viaduct and Bridge #6. A travel time of 72–77 min therefore yields an *average* speed for the flood wave of 0.32 to 0.33 km/min (~12 mph). Of course the actual flow speeds varied with location and the quantity of transported debris, and the formation and breakup of transient debris jams. The average hydraulic gradient, or energy slope, of the rivers from the dam to Johnstown is (471–353 m)/23.5 km = 0.005.

Some observers reported that Lake Conemaugh drained in as little as 37 min to as much as 75 min (McCullough 1968 p 102; Kaktins et al. 2013). John Parke clearly stated (Francis et al. 1891 p 451), "I do not know the actual time [to drain the lake through the breach], but it was fully forty-five minutes." Some modern compilations of dam breaches (Singh 1996; Froehlich 2008) cite the 45-min drain time but do not report it as a *minimum* estimate by Parke. Wahl's (1998) report transposed numbers, incorrectly noting that the breach took 45 min to fully form and the lake took 3.5 h. to drain.

To clarify, the historic records show that the South Fork Dam sustained overtopping for >3.3 h before catastrophically breaching in a very short time. The 1889 observations along with the hydrographs give strong support for the entire breach forming quickly. *At least 65 min were needed* for the lake stage to drop enough to expose the base of the upper breach. This may have led some observers to think the lake had drained in that time. However, the flow depth would still have been ~8 m in the lower breach. Certainly after 65 min large areas of former lake bottom would have been exposed by the rapidly falling lake level. It is therefore easy to see why different observers gave a wide range of drain times. The time to drain the lake would have been extended if the lowest part of the inner breach took a significant time to erode. One further insight from the hydrographs is that tailwater effects at the outlet were probably very small, otherwise even more time would have been needed to drain the lake.

As shown at lower right in Fig. 10.8, at late times the lake stage would have approached equilibrium as breach outflow dropped to near the rate of lake inflow. For example, if the inflow rate fell 40%, from $216 \, \text{m}^3 \, \text{s}^{-1}$ before overtopping (around 11:30 a.m., per Parke) to $130 \, \text{m}^3 \, \text{s}^{-1}$ several hours later, the reduced inflow could have sustained a flow depth in the lower breach >3 m. So even with a large drop in stream discharge the implied flow depth would be greater than estimated by Parke, who commented that when the lake drained there still remained in its bed a stream 4 or 5 feet deep (Francis et al. 1891 p 451). Parke had also added that "…this stream in the bed of the lake showed no signs of diminishing in volume until late in the following day, and was impassable with a boat for several days." No doubt the stream remained high for some time, as often seen in the receding limb of runoff hydrographs (e.g., see Fig. 9.4), but the dam was only threatened by the magnitude of the shorter-lived peak discharge.

The tendency of floods to recede relatively quickly in this "flashy" watershed is well illustrated by the behavior of the flood hydrograph from the 1977 Johnstown flood, which resulted from greater rainfall amounts than the 1889 flood. As measured at the USGS gage at East Conemaugh (see Fig. A3.1 of Appendix 3) this hydrograph on the main stem of the Little Conemaugh peaked at $1100 \, \text{m}^3 \, \text{s}^{-1}$ (Brua 1978). The discharge then *fell by 50% in only a few hours.* In this extreme runoff event, the discharge rate at this gage downstream from South Fork on the Little Conemaugh exceeded 50% of its peak for only 5 h. We note there were some differences in proportionate land use between 1889 and 1977, including surface mining, which would have contributed to the watershed response rate. That mining technique did not exist in the South Fork drainage at that time, but there were large areas that had been deforested for agriculture that would have contributed to rapid runoff. Also, the watershed response in 1889 for the South Fork of the Little Conemaugh should have been rapid given its much smaller drainage area than the combined basin area above the present East Conemaugh gage.

References

Brua SA (1978) Floods of July 19–20, 1977 in the Johnstown area, Western Pennsylvania. U.S. geological survey open file report 78–963, 62 p

Chow VT (ed) (1964) Handbook of applied hydrology. McGraw-Hill Book Co., New York, p 1418

Coleman NM, Kaktins U, Wojno S (2016) Dam-breach hydrology of the Johnstown flood of 1889 – challenging the findings of the 1891 investigation report. Heliyon 2, 54 p, https://doi.org/10.1016/j.heliyon.2016.e00120

Davis C (1889) The South fork dam, proceedings of engineers' Society of Western Pennsylvania, pp 89–99

Dingman SL (1984) Fluvial hydrology, W. H. Freeman & Co., NY, 383

Eng. News (1889) Engineering News, June 15 1889 p 541

Francis JB, Worthen WE, Becker MJ, Fteley A (1891) Report of the Committee on the cause of the failure of the South Fork dam. ASCE Trans. vol XXIV, pp 431–469

Frank WS (1988) The cause of the Johnstown flood: A new look at the historic Johnstown flood of 1889. Civil Engineering—ASCE, vol 58, no. 5, May 1988, pp 63–66. Also see: http://www.smoter.com/flooddam/johnstow.htm. [Accessed 2 Apr 2018]

Froehlich DC (1995) Peak outflow from breached embankment dam. J Water Resour Plan Manag 121(1):90–97

Froehlich DC (2008) Embankment dam breach parameters and their uncertainties,J Hydraul Eng, vol 134, no. 12, Dec 1, 1708–1721

Kaktins U, Davis Todd C, Wojno S, Coleman NM (2013) Revisiting the timing and events leading to and causing the Johnstown flood of 1889. PA Hist: J Mid-Atlantic Stud, vol 80, no. 3, 335–363

MacDonald TC, Langridge-Monopolis J (1984) Breaching characteristics of dam failures. J Hydraul Eng 110:567–586

McCullough D (1968) The Johnstown flood. Simon and Schuster, New York, 302 p

NOAA (2012) The enhanced Fujita scale, NOAA's Storm Prediction Center, Available at: http://www.spc.noaa.gov/efscale/. [Accessed 2 Apr 2018]

Pagan AF (1974) The Johnstown flood revisited. Civil Eng.—ASCE, Vol 44, No. 8, Aug 1974, pp 60–62

Pierce MW, Thornton CI, Abt SR (2010) Predicting peak outflow from breach embankment dams, prepared for National dam Safety Review Board Steering Committee on dam breach equations, Colorado state U, Ft. Collins, p 45

Pittsburgh Commercial Gazette (1889) Resident engineer Parke's story, Tuesday Jun 4th, p 1

PRR (Pennsylvania Railroad) (1889) Testimony Taken by the PRR Following the Johnstown Flood of 1889. [Statements of PRR employees and others in reference to the disaster to the passenger trains at Johnstown, taken by John H. Hampton, at his office in Pittsburgh, by request of Superintendent Robert Pitcairn; beginning July 15th, 1889.] Copy in archive of Johnstown Area Heritage Association. Many stories available online at: https://www.nps.gov/jofl/learn/historyculture/stories.htm [Accessed 1 Apr 2018]

Russell TH (1964) All at once the dam was gone, Johnstown Tribune-Democrat, May 29

Singh VP (1996) Dam breach modeling technology. Kluwer Academic Publishers, Boston, p 260

Tracy HJ (1957) Discharge characteristics of broad-crested weirs USGS Circular 397, p 15

Unrau HD (1980) Historic structure report: the south fork dam historical data, Johnstown flood National Memorial, Pennsylvania. Package no. 124, US Dept of the interior, NPS (1980): 242

Wahl TL (1998) Prediction of embankment dam breach parameters - a literature review and needs assessment. DSO-98-004, dam safety research report. Dam safety office, water resources res. Lab., p 59

Walder JS, O'Connor JE (1997) Methods for predicting peak discharge of floods caused by failure of natural and constructed earthen dams. Water Res Research 33(10):2337–2348

Wellington AM, Burt FB (1889) "The south fork dam and Johnstown disaster," Eng. News and Railway Journal 21:540–544

Chapter 11
Discussion

11.1 Suppression of the Investigation Report

Even in the politics of today, our Congress often applies one tactic to deal with controversial issues: delay, delay, delay! This same tactic was used to soften the impact of the ASCE report and also to allow time for President Becker to edit its content to make it more palatable. The Club members had limited control over the hydraulic engineers on the Committee, so their best procedure was to delay the report if they could. In this they succeeded royally, as it did not see the light of day until late May, 1891 in Chattanooga, TN, far from Johnstown, and 2 years after the flood.

The delay seems to have begun in New York. ASCE had its regularly scheduled meeting there a few days after the Johnstown Flood. A member from Pittsburgh spoke at some length about the South Fork dam and its history, with details that could be quickly gathered. The speaker was William Shinn, who was not a hydraulic or dam engineer, but happened to be a former managing partner of Andrew Carnegie. Shinn was not a member of the South Fork Fishing & Hunting Club, but was a member with Carnegie, Frick, and Pitcairn of the exclusive Duquesne Club in Pittsburgh. Shinn's presence at the New York meeting was not otherwise unusual as he was one of ASCE's directors at that time. He had left Pittsburgh on Tuesday, June 4th, and had to travel by roundabout rails because the PRR line to Altoona was closed with bridges and embankments under repair. But Shinn was indeed present, wanting very much to be there. At that meeting the committee members were chosen to investigate the South Fork dam breach. Three preeminent hydraulic engineers were selected, but then President Max Becker, a railroad man with virtually no dam or hydraulic engineering experience, was inserted to chair the committee. Unlike the hydraulic engineers, Becker was susceptible to possible influence. He served as chief engineer of a railway that was controlled by Pitcairn's Pennsylvania Railroad, via stock ownership. And like most of the Club members, Becker lived in Pittsburgh.

© Springer International Publishing AG, part of Springer Nature 2019
N. M. Coleman, *Johnstown's Flood of 1889*,
https://doi.org/10.1007/978-3-319-95216-1_11

At the outset of the investigation Vice President Alphonse Fteley publically stated his views about the need to quickly release the report. The committee had just returned from visiting the ruined dam, and the members were attending the ASCE annual convention in Seabright, NJ. They had climbed over the dam remnants and gathered their own survey measurements. On June 22nd Fteley gave a preliminary report about Johnstown. He felt it premature to state causes about the dam break, but he added there was no doubt the disaster would not have happened if the spillway had adequate capacity. Fteley went on to say "It is very essential that the truth of this matter should be known, as it may show that other dams are unsafe in the light of recent experience." [*Eng. News* 1889 p 603] By his own words Fteley wanted the facts to come out and would have opposed keeping the report secret until 2 years after the flood.

The report they developed was much longer and more detailed than ASCE's previous report on a dam failure, the Mill River disaster in 1874. I have found no information about who marshalled the drafting of the South Fork report and which of the engineers prepared its various parts. James Francis was dealing with the death of his daughter-in-law Caroline on July 30, and the following day Alphonse Fteley left for a month's vacation in Europe. In October, as the investigation continued, William Shinn was nominated to become the new ASCE president. At that time, being nominated virtually assured election to the office. The report was completed and submitted on January 15, 1890, 7 months after the flood. Although signed and submitted the report was not released. It was "sealed" and given to Max Becker. That was unusual in itself, as reports were normally kept by the Secretary. Becker then turned over the reins of the presidency to William Shinn. As President, Shinn could have immediately directed that the report be made public, but Becker was given full authority to decide the appropriate time to release the report. This seems to me a serious mistake – the decision about release should have been placed in the hands of the safety and subject-matter experts, the hydraulic and dam engineers. What truly happened to this report after it was "sealed?"

The ASCE membership widely believed that the South Fork report would come out during the 1890 meeting in Cresson, PA, a year after the flood. But with Shinn as president, and prominent railroad men on the organizing committee (i.e., Club member and PRR executive Robert Pitcairn and his assistant Michael Trump), the report was discussed in secret and not released. Alphonse Fteley came to the Cresson meeting but William Worthen stayed away. Both Becker and James Francis spoke to reporters at Cresson. Becker declared that to keep the ASCE out of ongoing lawsuits the report would not be released. Francis also noted this but said it was time for the report to come out, that it would be released soon. Unknown to him, yet another year would pass before its release. A rail excursion to Johnstown was scheduled for ASCE attendees and their wives, but there were no visits to the South Fork dam as part of the meeting. Some engineers went there on their own.

William Shinn and Max Becker kept the investigation report under wraps through the term of Shinn's presidency. Octave Chanute was elected ASCE president for 1891 and still the report remained secret. He was retired at that time and his income mainly depended on consultant fees with the railroads. Although I found no written evidence of this, Chanute might have been pressured to leave the fate of the South Fork report in the hands of former president Max Becker. In any event, the sealed report was in Becker's hands to handle as he saw fit.

The report at last was presented during the ASCE convention in Chattanooga, TN. The membership surely saw that something was afoot. Of the four committee members, only James Francis attended to give the long-awaited findings. The membership must have discussed among themselves why Worthen, Fteley, and Becker were not present. No doubt the members of the South Fork Fishing & Hunting Club would have approved of the Tennessee location for releasing the report. Chattanooga was far from Johnstown and South Fork, so no one could hear the findings and then conveniently visit the ruined dam to judge for themselves the merit of the report. And by that time, 2 years had passed since the dam breach. Rain and frost and the footsteps of hundreds of sightseers would have altered to varying degrees the surfaces of the dam remnants.

11.2   Review of the ASCE Report (Francis et al. 1891)

I respectfully disagree with the ASCE report on the cause of the failure of the South Fork dam (Francis et al. 1891). The committee had concluded that the modifications to the dam by the Club "…cannot be deemed to be the cause of the late disaster as we find that the embankment would have been overflowed and the breach formed if the changes had not been made." The report's conclusions are questionable for five reasons:

First, if the drainage pipes had been replaced they would have been opened early on the day of the flood to their maximum capacity, which would have reduced the rate of rise observed by John Parke. The issues noted in Chap. 3 about possible hydrodynamic damage to the pipes and culvert would not have been a concern, because releases from the dam were no longer needed to regularly supply water to a canal. Therefore repaired or newly installed pipes and a repaired culvert would only have been used during floods or to reduce lake levels to allow for embankment repairs.

Second, the committee failed to recognize that the dam likely had been lowered by the Club more than 0.6 m to at least 0.9 m (3 ft), and possibly more in the center due to settling of the repaired embankment. The volume-elevation curve (Fig. 9.3) for Lake Conemaugh shows that an added 0.9 m of lake stage would have stored an additional 1.6 million m^3 of water without overtopping.

Third, in the original design the main spillway would have had much greater discharge capacity, and the discharge pipes and auxiliary spillway would have been functional. But the committee never acknowledged that the auxiliary spillway existed, even though they took their own survey measurements of the dam remnants.

Fourth, contrary to the committee's claims of an increasing rate of rise of the lake stage until the time of dam failure, and beyond, there is evidence that flood levels in both the South Fork and the Little Conemaugh had stabilized and begun to fall, if only slightly, by 12:00–1:00 pm, suggesting that the peak discharge in the runoff hydrograph occurred hours before the dam breach. Most of the observational evidence about the streams was contained in the Pennsylvania Railroad testimonies (PRR 1889). In support of the ASCE engineers, I found no proof that the PRR gave those testimonies to the investigation committee. Those testimonies were recorded 4 to 5 months before the ASCE report was completed. The stream observations would have been very helpful to the committee because they were contrary to an assumption in their report that river levels would have continued to rise even after the dam breach.

And fifth, the poor reconstruction work by the Club had several adverse consequences. Along with the fatal lowering of the dam crest, the random-fill technique used to repair the partial breach of 1862 (rather than the clay "puddling" method) was a major factor in the disaster. The random fill made it possible for water to penetrate a large portion of the dam's core. The penetrating water could saturate and destabilize large volumes of the embankment during high lake stands. The committee itself reported (Francis et al. 1891, p 454) that "All the material put in [by the club]…to repair the breach of 1862, appears to have been washed out, together with part of the old embankment…" They also wrote (p. 454) that exposed parts of the *original* dam showed that they "…offered great resistance to washing and that [the work] was originally selected and put in with the requisite care to make a sound embankment."

Figure 11.1 reveals how the dam *as originally built* with a higher crest would have resisted overtopping on May 31, 1889. Even if extremely high lake inflows had continued unabated (an unrealistic assumption), overtopping of the dam at its design height would have been averted for ~14 hrs. Without another failure mechanism the dam would have been preserved, because lake inflows would have substantially diminished during the afternoon and evening. In the actual event, if the peak stream inflow had time to fall just 9% (from ~216 to ≤197 m^3 s^{-1}), overtopping would have been entirely prevented if the embankment had not been lowered by the Club. Fig. 11.1 also reveals that if the reconstructed dam had been built only 0.6 m higher, overtopping would have been delayed more than 7 h.

It is a little known fact that the reliability of the South Fork dam was tested just a few years after it was completed. In the spring of 1856 the reservoir overflowed at an unspecified place after a rapid snowmelt (Unrau 1980 p 51). I have not found a description of where the overflow occurred, but given the higher dam crest it must have flowed through the auxiliary spillway on the southwest abutment. The reservoir appears to have performed as intended, with two functioning spillways and the discharge pipes also probably opened. A leak in the dam was reported at that time but was soon repaired. In fact, the Pennsylvania State Engineer inspected the South Fork dam later that year: "The Western Reservoir [Lake Conemaugh] was examined and found to be in excellent condition. It furnished a sufficient supply of water to keep up the [canal] navigation when other sources had entirely failed" (Gay 1856, p. 16). He had reported that the spring floods severely damaged other impoundments in the region, including Piper's and Raystown dams on the Juniata River. Eighty-foot breaches occurred in both of these dams.

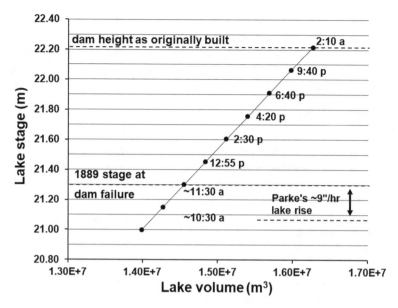

Fig. 11.1 Graph showing time needed to reach various hypothetical lake stages. Graph was prepared using the discharge capacity curves in Fig. 10.4 with the storage elevation curve in Fig. 9.3. Plotted times assume that a constant lake inflow of ~216 m³ s⁻¹ would continue for *many* hours. Even in this extreme and unrealistic scenario, the nonlinear increase in spillway and pipe flow capacities and lake storage would have averted overtopping for ~14 h. In reality the dam would never have overtopped

Useful insights can also be gained from the resilience of another earthen dam in the region that was built with a "puddled" clay core. This dam was on Mill Creek four miles from Johnstown, and it survived the May 1889 event. We know this from the minutes of the Board of Trade and Citizens Meeting, June 19, 1891, which state that the dam was 310 ft. long and 25 ft. high, built in 1834 with a spillway 44 ft. wide and a "puddle wall" 7.5 ft. wide within the center of the dam. The minutes go on to document that a "freshet" [flood] occurred on July 2, 1889 and that the spillway was insufficient to carry the flow. The *whole dam* was overtopped by this flood but it did not fail, providing an example of the durability of the "puddle wall" construction technique. Unfortunately, the South Fork Fishing & Hunting Club did not use this technique to repair the South Fork dam. Saturated conditions would readily have penetrated to the center of the dam, greatly reducing its shear strength. Therefore, liquefaction effects may have been largely responsible for the very rapid upper breach formation. Francis et al. (1891 p 446) commented that hauling fill by [horse] teams over the freshly deposited material "…made a fairly compact embankment on the upper side of the stone embankment." I find it hard to believe that eminent engineers like Francis and Worthen would fail to comment about the lack of "puddle" ditches, a standard technique at that time, or that they would infer that horse hooves and cart wheels might achieve similar results. Since there was a definite sag in the central, filled portion of the embankment, any compaction from horse teams was limited.

It is interesting to compare the committee's conclusions about the South Fork dam to their earlier investigation of the Mill River dam failure in Massachusetts (Francis et al., 1874). That dam failed on May 16, 1874, killing 139 people. A remarkable account of the Mill River disaster and the investigation that followed was written by Elizabeth Sharpe (2004). James Francis and William Worthen served on the committees to investigate both dam breaks. Their report on Mill River was rapidly published, adopted on June 10th, less than 1 month after the disaster. They criticized the material used to make the Mill River embankment, and that it could not be relied on to make the structure water-tight. During the construction there "… was no sufficient inspection, so peculiarly important in a work of this description…" "The remains of the dam indicate defects of workmanship of the grossest character." In the discussion section, Worthen went on to write:

> Men were employed who were ignorant of the work to be done, and there was nothing like an inspection, although money and life depended upon it. I do not believe, however much we are an evolved species, that we are derived from beavers; a man cannot make a dam by instinct or intuition.

Neither Worthen nor Francis reflect this philosophy in their role 16 years later as investigators of the South Fork dam failure. The need for engineering inspection and water-tight embankments was equally important during the Club's repair of the South Fork dam. And yet, when confronted with the many changes made to that dam and the poor methods used to repair the embankment, including lack of engineering inspection, the South Fork report concluded that the changes and repairs to the dam were not responsible for the disaster. They even went so far as to conclude that if the embankment had been rebuilt to its original height the result may have been *greater* loss of life. Clearly there were inconsistencies in philosophy between the ASCE reviews of the South Fork and Mill River dam failures. The views of Worthen and Francis would not have changed over the years. It is likely that President Becker removed from the report any concerns about the lack of engineering inspection in the Club's dam work, perhaps at the urging of Pitcairn.

One further point should be mentioned. The three-member committee that investigated Mill River included only hydraulic engineers, eminently qualified for that review. The ASCE president that year, Julius Walker Adams,[1] was not among them. But for the South Fork review 15 years later the standing President, Max Becker, was inserted on the committee even though he had virtually no experience with dams. He was a railroad man who was given complete control on when to release the report, even after he was no longer President. That decision should have been entirely in the hands of the subject matter experts, the hydraulic engineers.

Although Francis et al. (1891) found the Club's mode of repairing the breach of 1862 was not according to best practice, they nonetheless concluded that "…failure of the dam cannot be attributed to any defect in its construction. The failure was due to the flow of water over the top of the earthen embankment, caused by the insufficiency of the wasteway…" Their flow calculations would have been quite different

---

[1] Adams was primarily a railroad man but also had experience in large scale water and sewer projects.

had they included the auxiliary spillway, but their report makes no mention of it. Our calculations (Coleman et al. 2016) show that the changes made to the South Fork dam were indeed responsible for the disaster, having altered its original design and rendered it fatally vulnerable to overtopping. The dam could indeed have survived the rainfall event of May 30–31, 1889 had it been maintained as built in 1853 with a higher crest, a functioning second spillway, five drainage pipes, proper well-compacted fill with puddle layers, riprap replacement of proper size, and no bridge or fish screens across the main spillway.

The discharge capacity of the original dam was more than *twice* that of the reconstructed dam. The dam as originally built could have averted overtopping until the next day, even under extreme conditions of inflow duration and rate. Stream flows peaked hours before the dam breach. Flood levels would have rapidly declined, saving the dam and the towns below it. I therefore disagree with the main conclusion of Francis et al. (1891), that the dam failure was inevitable. They placed too much reliance on estimates for the rate of rise of the South Fork and the implied excess flow into the reservoir. This seems unusual given James Francis' expertise in flood control and weir calculations and his practical knowledge of river behavior in flood, from the rapid rise to peak discharge and the relative speed of flood recessions. But as noted earlier, Francis and Worthen had long experience with the Merrimack River, which has a much larger drainage basin and protracted flood peaks.

The committee assumed flow depths in the spillway that were implausible, based on evidence from the dam remnants that they themselves visited (i.e., preserved plow furrow on crest). Their unsupported assumption of protracted, extreme lake inflow before and after the dam breach led to their conclusion about the non-survivability of the dam.

11.3 Whitewash and Distortion of the ASCE Report

There are distinct lines of evidence that the report by Francis et al. was not only delayed but watered down and whitewashed. Key facts were excluded and some things were added that had no relevance to the dam breach but were related to PRR liability. Especially telling – the report does not criticize the Club for failing to involve qualified engineers in the dam work. Certainly the Club men were sensitive about this issue which related to liability claims. The Club's failure to involve experts had been widely broadcast in the engineering periodicals in 1889, and was clearly expressed in ASCE's earlier review of the Mill River dam in 1874. Both Francis and Worthen were part of that review. They would have had the same concern for the South Fork dam, but it clearly was edited out, excluded from the report. Pitcairn himself in his PRR testimony declared that the late Benjamin Ruff, who had supervised the repairs for the Club, was "…better than any engineer." But Ruff's experience with railroad embankments was no substitute for expertise in dams and canals. He was not better than any engineer, and was woefully inept when compared to experienced hydraulic engineers. I will suggest that Pitcairn used his influence over Max Becker to modify the report. But when the report was eventually published, the glaring omission to critique the Club about engineering expertise told the professional community that all was not right with the investigation.

The committee analyzed the spillway flows to compare the discharge capacity of the original dam vs. the Club's modified dam. They used their own survey measurements to show that the crest of the dam had been lowered by ~2.04 ft. (0.62 m). But whoever did this calculation distorted the result by including a high data point on the natural surface of the eastern abutment. It was not part of the constructed embankment. If that data point had been left out, the amount of lowering would have been significantly greater. And, if I may say right here, averaging the data to estimate lowering of the crest was a bit of a sham. The lowest point on the dam crest, not the average, determined the overtopping level. The committee should have used their own data from the eastern dam remnant to interpret the low points and degree of lowering. Station 9 confirms the crest was lowered 3 ft. (0.9 m), and Station 8 + 50 shows an even greater lowering of 3.43 feet, (>1 m). Both stations were far enough from the breach margin that they were not likely changed by the dam rupture. Bottom line – the committee's calculations were distorted by averaging and including an inappropriate data point, thereby overstating the discharge capacity of the Club's modified dam.

The Committee report repeated the old story that the dam crest was lowered to widen the road. They did not point out the obvious hazard of doing so, thereby reducing the spillway capacity. The eminent hydraulic engineers would have immediately pointed this out but must have been overruled by Becker. Nor did the report point out in civil engineering terms that a road atop a dam can simply be widened by adding construction material along the margins of the crest, then paving the surface with gravel. The hydraulic engineers surely saw that the Club's true motive was to lower the surviving dam remnants so that fewer tons of fill would be needed to fill the breach to the level of the new lowered crest. If they criticized the Club for this in a draft report it must have been edited out by President Becker.

The voluminous testimony gathered by the Pennsylvania Railroad after the flood was apparently not given to the ASCE investigators. For example, they could have considered observations about local stream levels. Their extreme calculations of lake inflows would have been tempered by reports that the rivers had ceased to rise around noon and may even have begun to fall.

The PRR testimony also chronicled the fate of people on the Day Express trains. Someone, likely Pitcairn via Becker, injected text into the ASCE report (p 434) that said 28 passengers "…were drowned in the attempt to save themselves by running from the train to the higher ground on the north side of the tracks." This appeared designed to protect the PRR from liability by implying the passengers "killed" themselves – they should have stayed on the train. But in fact the PRR testimony showed that passengers on one of the trains had been forewarned about the South Fork dam. They had the farthest to go to reach safety, but were ready to run when the wave came and escaped. The PRR may then have been liable for not warning everyone of the danger they knew about the dam that day. Pitcairn's railroad testimony falsely claimed that the passengers and crew at East Conemaugh had been "thoroughly posted" about the dam. Other text in the investigation report was also designed to protect the PRR. Page 434 of Francis et al. (1891) stated that further warnings about the dam would have saved few lives in Johnstown because the streets were already flooded, trapping folks in their homes. There was truth in that, but more people would have waded or rafted to safety had they gotten warnings directly from top railroad executives like Pitcairn and Trump. And what of the towns all along the flood path, west of South Fork, and the many PRR workers in East Conemaugh? Few received the warnings.

The most puzzling omission of the ASCE report was failure to perceive an auxiliary spillway on the western abutment. Part of that abutment was clearly excavated to a relatively flat surface. The width of the lowest portion when added to the width of the main spillway was ~140 ft. or more, approaching the 150 ft. width in the original plans by the state engineer. It is inconceivable that hydraulic engineers of the caliber of Francis, Worthen, and Fteley, who had personally walked and surveyed the remnants of the dam, would have overlooked the second spillway, the very place where a ditch was desperately dug on the day of the flood to try to save the dam. They had measured elevations to 1/100th of a foot. At the dam's original design height, flows 3 feet deep would have crossed the western abutment before the crest overtopped.

The evidence of an auxiliary spillway may have been stricken from the report by Becker, using his authority as President. But regardless of ASCE protocols at that time and what may have been blind deference to President Becker, the hydraulic engineers should have stood their ground on scientific principles. Perhaps James Francis was speaking to us down through the years. In his solo presentation at Chattanooga, Francis commented that "...near the ends [of the dam] there were ascents to the level of the top of the dam." That evidence for an auxiliary spillway became part of the discussion section in the report, and could not be edited out by the absent Becker (Francis et al. 1891 p 468).

I recognize that the late 1800's were a different era, with gentlemanly behavior and decorum being important in professional business and society. Decorum may have been more essential then than now. But far too much deference was apparently given to Becker to decide when the report could come out and what contents were acceptable. Francis' own words show his concern about the report being withheld, even a year before its actual release. A friend of Francis' described him as a man who sought to avoid arguments and went about his own business. The hydraulic engineers were not susceptible to strong-arm tactics by the railroad men. Francis in particular was comfortably retired with no financial worries, but by the time of the report's release the illness that would claim his life had likely begun to affect him. He would die within 15 months. Worthen and Fteley clearly had concerns about the whole business because they stayed away from Chattanooga, where the report finally came out. Could it be that their "signed and sealed" report had been edited after the fact without their knowledge? There are so many discrepancies in the report that this scenario must be considered.

The report as finally published contains only one criticism of the South Fork Fishing & Hunting Club. Even that was turned into a positive for the Club, which undoubtedly pleased its members. Francis et al. (1891) said that the changes made to the dam by the Club may have caused it to fail earlier in the day compared to the higher, original dam, but the dam would have failed even had it not been lowered, and "…the result would have been equally disastrous, and possibly even more so, as the volume of water impounded was less…" The messages embedded here were that the Club's changes did no real harm because the dam built by the State would also have failed, and the inevitable flood was less severe because the Club lowered the dam. But this claim was not true.

In brief, the ASCE report was skillfully edited to favor both the Club and the Pennsylvania Railroad. Pitcairn's fingerprints are figuratively all over the document. Messaging by the Club was not restricted to influencing the content of the report. At least two post-flood letters were sent to the *Johnstown Daily Tribune*, trumpeting the warnings that had been sent to Johnstown. One was from Joe Wilson, superintendent at the Argyle Mines in South Fork. Wilson wrote that he twice sent Dan Siebert to check on the South Fork dam and sent warnings that were heeded in South Fork. Wilson claimed to have telegraphed: "The dam is breaking. Look out!"

The other letter came from Boyer, supervisor of grounds at the Club. Boyer described the summer resort, including its "fine clubhouse" with 47 bedrooms and amenities. He then wrote that "…all parties between the lake and Johnstown were notified three hours in advance of the flood that there was great danger of the dam breaking, and that it probably would, and warning them to flee. But they would not heed." That letter drew a quick reply (June 26th) to the *Tribune* from T. P. Williams, who was also responding to quotes from sermons "all over the land" about unheeded warnings. Williams wrote that for 4 h before the disaster he was near the Pennsylvania Railroad, and for 2 of those hours was on a roof in earshot of the railway and near the telegraph tower. But the only warning he got was "…the Lincoln bridge followed by a train of [rail] cars, floating toward us, mowing our homes down before our eyes… Heaven knows our burdens are enough to endure, our losses enough to sustain, without the publication of such reproachful remarks, as these misinformed people are pleased to indulge in."

11.4   Who should be Blamed for the Johnstown Flood?

The organizers of the South Fork Fishing & Hunting Club were responsible for acquiring the property and making repairs to the dam. The resort was the brainchild of Benjamin Ruff (Fig. 11.2), who oversaw those repairs and chose not to involve qualified engineers at any stage. He died in Pittsburgh on March 29, 1887, 2 years before the 1889 flood. Therefore, Ruff never knew that his changes to the dam would, in the end, destroy thousands of people who lived and worked below the dam. Ruff had rejected the sound engineering comments by engineer and geologist John Fulton. I conclude that the bulk of responsibility for the dam failure rests with Ruff. Had this disaster happened present-day, the Club would certainly have been liable for the loss of lives and property. Interestingly, the greatest changes to

Fig. 11.2 Portrait of
Benjamin Ruff

Pennsylvania's laws pertaining to dams and levees did not happen until the Laurel Run dam failed in 1977.

The death toll from the Johnstown Flood would have been reduced if Michael Trump or Robert Pitcairn had sent warnings *under their names* about the condition of the South Fork dam. They were onboard eastbound trains and had warnings in hand about the danger to the dam, and Pitcairn himself had telegrams from people he most trusted, an early message from Club president Unger who was at the dam, and Joe Wilson. He had asked Wilson in particular to keep an eye on the dam for him. Telegraph lines were open to Johnstown and beyond East Conemaugh, and the telephone exchange in Johnstown was still operating. A direct warning from Pitcairn himself, whose name was well known in Johnstown, would have carried great weight, and no doubt many more people would have braved the water in the streets to reach higher ground. Fewer would have died. And no less than the people of Johnstown, Pitcairn and Trump should have warned the stranded passengers on three trains and their own employees who were in East Conemaugh and were among the first to be struck by the flood wave. Pitcairn could not have known the extraordinary damage that awaited Johnstown, but as a Club member who knew the size of Lake Conemaugh, he had to have understood the risk to people in the narrow valley of the Little Conemaugh River.

Robert Pitcairn was at the nexus on the day of the flood. Except for Michael Trump, no one else knew what he knew. He was informed of severe problems from flooding along the line, had the early warning from President Unger at the dam, as a Club member he knew the size of Lake Conemaugh, as superintendent had full control of the railroad crews, and rapid telegraphic communications at his fingertips. As an executive, he could make rapid decisions about ticket prices, train schedules, freight movements, and probably also the hiring and firing of people. But when lives were on the line on May 31st, Robert Pitcairn hesitated for hours and never alerted the public with a "message from the Superintendent." He must bear the historic blame for that.

The South Fork dam was properly built with spillways at both ends, as intended by state engineer William Morris. He had also designed the eastern dam at Hollidaysburg, which had a similar spillway width (see Chap. 13). However, the South Fork auxiliary spillway was less than half as deep as the curving spillway on the northeast end. The years of delays in funding and completing the dam were no doubt responsible for this engineering shortcut. Even so, calculations show that the combined action of the two spillways and the availability of five large discharge pipes would have had enough flow capacity to prevent overtopping on May 31, 1889. The dam would have been saved along with thousands of people.

A recent historical perspective has been written about the various owners of the South Fork dam. Kooser (2013) asserts that the four successive owners of the South Fork Dam were all to some degree responsible for the dam failure. The "owners" included the Commonwealth of Pennsylvania, the Pennsylvania Railroad Company, John Reilly (congressman), and finally the South Fork Fishing and Hunting Club. But the Pennsylvania Railroad had simply acquired the Main Line Canal system from the State, along with the two large dams at South Fork and Hollidaysburg. A partial breach of the South Fork dam happened under PRR ownership, but caused little damage. The PRR left the dam in that condition, posing no risk to anyone downstream. No fault can be attached to the railroad. And John Reilly was not an engineer. He never attempted to repair the South Fork dam and simply sold the property to the Club. It is also unclear whether Benjamin Ruff shared all his plans for the property with Reilly. Reilly was not at fault. Kooser (2013) also claims that the dam was poorly designed, built, and maintained long before the Club's ownership. He criticizes the earthen embankment design, but this is the most common form of dam in the world. Overtopping of such dams must be prevented by the design of adequate spillways. Hydraulic calculations (Coleman et al. 2016) reveal that the South Fork dam as originally built would indeed have avoided overtopping and survived the storm of May, 1889.

Should the Club members be viewed as personally responsible for the disaster? They have long been blamed. They were successful businessmen, some with extraordinary wealth, but with little or no knowledge of dam safety or hydraulic engineering. In reality they had no more interest in the infrastructure of their Club than members of a country club today would worry themselves about maintenance of fences, water hazards, landscaping, or the health of the club's trees. The South Fork Club members simply wanted an exclusive, quiet place to relax and escape the smoke and soot, summer heat, and epidemics of Pittsburgh, and a safe place where their wives and privileged children could meet others of their class. The bar and porch of the Clubhouse were places for the men to ponder current and future business ventures.

Nonetheless, the Club members must bear their share of blame for the Johnstown flood, the reason being the membership of Daniel Morrell. In late 1880 he had sent to Benjamin Ruff an engineering report by Robert Fulton (1880), outlining problems with the South Fork dam. Ruff rebuffed those comments in his letter reply on

Dec. 2, and also seems to have ignored a follow-up letter (Dec. 22) from Morrell, which offered assistance to improve the dam. It is possible that Ruff did not share the contents of these letters with Club members. Even if that was true the members soon learned of Morrell's concerns. As the General Manager of Cambria Iron and a leading citizen of Johnstown, Morrell had every reason to want a large dam above the town to be solid and safe. So he applied for membership in the Club and was approved. Morrell was a giant in the steel industry, and just as Andrew Carnegie was a household name in Johnstown, Morrell was a household name in Pittsburgh where most of the Club members lived. He was a businessman like themselves. And once he was a Club member he had social opportunities to express apprehensions about the dam, as Pitcairn himself acknowledged. Morrell had a sterling professional reputation, and any concerns of his should have raised some "red flags" at the Clubhouse. *The members clearly knew of valid concerns about their dam.* Just as Ruff died 2 years before the flood, Daniel Morrell likewise did not live to see the disaster. He had been in failing health for several years, passing away on August 20, 1885, at age 64. But he had lived long enough to verbally express his anxieties about the dam to Club members and officers, consistent with his earlier letters to Benjamin Ruff. Unfortunately, the Club never acted on his concerns.

Also, the wealthiest members of the South Fork Fishing & Hunting Club could and should have done *much more* to help survivors of the flood. Although some members did contribute, their offerings were small compared to the large sums and donations from many U.S. cities and from Europe. More than $3.7 million had been contributed, not including food and supplies. Even school children and convicts sent coins (McCullough 1968), which truly makes the wealthier Club members seem a stern and miserly bunch. I believe the Club members did not offer nearly enough help after the flood, a fact that history should always remember.

The 1889 breach of the South Fork dam obliterated more than 2200 people, a majority of whom were women and children. The future reputations, historical legacies, and perhaps fortunes of the Club members were at stake. Tried and convicted in the press, several Club members sought to rescue those reputations by using their power to defeat legal challenges. Liability suits in Pittsburgh were ably defended by attorneys and Club members Knox and Reed. There can be little doubt that the juries hearing those cases would have included railroad and steel workers whose living wage could magically disappear unless a non-guilty verdict (or no verdict) resulted.

The Pittsburgh men covered their tracks well, but enough is now known to see their intentions. The Club members wanted to control the message in the formal ASCE engineering report, which they believed would carry great weight in the judgement of history. But the evidence shows that Robert Pitcairn, and through him other members, worked to suppress and whitewash the ASCE investigation, a report by some of the most prominent hydraulic engineers in the country. These actions to distort history are also a part of the Johnstown Flood legacy.

So a mystery remains, why would eminent hydraulic engineers of the caliber and integrity of Francis, Worthen, and Fteley have signed a safety report that contained information they had to know was incorrect or, at the very least, questionable.

Perhaps the "sealed" report was altered after they signed it, before publication. Whatever happened, Robert Pitcairn and Max Becker undoubtedly had a role. And what was the true role played by Becker? He personally delayed the report, but was he strong-armed into this by Pitcairn or did he willingly go along with his railroad colleague? Perhaps this was his response to the election controversy at the end of 1888, when some New York engineers tried to stop his election as President. Like Janus, Max Becker will be remembered here with two faces - we may never know his true motives. And if Francis, Worthen, and Fteley chose to appeal Becker's handling of their report to a higher power, the ASCE president in 1890 was now William Shinn, the former managing partner of Carnegie. The power of the Club maintained its grip. Shinn would simply have deferred to Becker.

I found no hard evidence that the "sealed" report was altered after the fact, but its contents and the motives of figures like Pitcairn and other Club members lead to that conclusion, or at the very least to severe editing before it was sealed. There may be clues in the surviving personal papers of these men that could help answer this. The records I reviewed about Francis did not, but his notes, notebooks, and correspondence are extensive. I must leave this mystery for others to ponder.

References

Coleman NM, Kaktins U, Wojno S (2016) Dam-breach hydrology of the Johnstown flood of 1889 – challenging the findings of the 1891 investigation report. Heliyon 2 (6), 54 p, https://doi.org/10.1016/j.heliyon.2016.e00120, e00120

Eng. News (1889) *Engineering News*, 6/29/1889 p 603

Francis JB, Ellis TG, Worthen WE (1874) The failure of the dam on Mill River. ASCE Trans. vol III, no. 4, p 118–122

Francis JB, Worthen WE, Becker MJ, Fteley A (1891) Report of the Committee on the cause of the failure of the Soth Fork dam. ASCE Trans. vol XXIV, Issue 477, p 431–469

Fulton R (1880) Letter report to DJ Morrell concerning South Fork dam, dated Nov 26 1880

Gay EF (1856) Report of the [Pennsylvania] state engineer for the fiscal year ending Nov. 30, 1856. Report dated Dec 23, 1856. In: reports of the heads of departments. Transmitted to Governor in pursuance of law, for financial year ending Nov 30 1856. Harrisburg

Kooser NM (2013) Determining responsibility for the failure of the South Fork Dam, Unpublished thesis, American Public University System, March 11 2013, Charles Town WV [copy on file at Johnstown Area Heritage Assoc., Johnstown, PA]

McCullough DG (1968) The Johnstown Flood. Simon & Schuster, New York, 302 p

PRR (Pennsylvania Railroad) (1889) Testimony Taken by the PRR Following the Johnstown Flood of 1889. [Statements of PRR employees and others in reference to the disaster to the passenger trains at Johnstown, taken by John H. Hampton, at his office in Pittsburgh, by request of Superintendent Robert Pitcairn; beginning July 15th, 1889.] Copy in archive of Johnstown Area Heritage Association. Many stories available online at: https://www.nps.gov/jofl/learn/historyculture/stories.htm [Accessed 1 Apr 2018]

Sharpe EM (2004) In the shadow of the dam – the aftermath of the Mill River flood of 1874. Free Press, New York, p 284

Unrau HD (1980) Historic structure report: the South Fork dam historical data, Johnstown Flood National Memorial, PA. Package no. 124, US Dept of Interior, NPS (1980):242

Chapter 12
Conclusions

It was a classic tale of power over truth, partly revealed by the power of the press, our precious gift from the founding fathers. Key members of the South Fork Fishing & Hunting Club had the motives and the power to influence the timing and the content of the investigation report. It was all done through the railroad men. The Club was fortunate in 1889 – ASCE's president was engineer and railroad man, Max Becker. He was the Chief Engineer of the Pittsburgh, Cincinnati & St. Louis Railway, a company controlled via stock ownership by the powerful Pennsylvania Railroad (PRR). His railway's eastward movements, passengers, and cargos had to be coordinated through Pittsburgh. And the Superintendent of the PRR's Pittsburgh division was the Club's own Robert Pitcairn, a boyhood and lifelong friend of Andrew Carnegie. Carnegie clearly distanced himself from the Johnstown flood, never mentioning the tragedy or his Club membership in an autobiography published after his death. Scores of pages from the Club's guestbook that would have confirmed his visits were torn out of the register.

Much has been written of warnings sent on the day of the flood to the people of Johnstown. During stormy weather they had grown accustomed to cautions about the dam that had never borne true. But the people had *never* been warned directly by Robert Pitcairn, a household name in Johnstown. He had in hand messages confirming a severe threat to the South Fork dam from people he could rely on, Unger and Wilson. The message from Unger came to him before he left Pittsburgh, hours before the flood struck Johnstown. Pitcairn bears responsibility for not sending an evacuation warning under his name and title to Johnstown and to his own employees and passengers in East Conemaugh. His rapid response after the flood to repair the railways is commendable, quickly bringing aid, but also was driven to restore commerce and profits for the railroad.

ASCE's investigation report was finished by January 1890. As Becker stepped down he sealed the report and handed the reins to a new president, William Shinn of Pittsburgh, a former managing partner of Carnegie's at the Edgar Thompson works. Shinn had been nominated to become president while the report was being finalized. The sealed report should have been placed in the hands of the Secretary, but Becker

© Springer International Publishing AG, part of Springer Nature 2019

N. M. Coleman, *Johnstown's Flood of 1889*,

https://doi.org/10.1007/978-3-319-95216-1_12

kept it. Together, Becker and Shinn kept the report sealed through the end of 1890. They may have thought that memories of the Johnstown flood would slowly begin to fade. The ASCE convention that summer was in Cresson, just east of South Fork. Pitcairn and his assistant Michael Trump were not ASCE members but had eased their way onto the organizing committee. The convention was held at the Mountain House, run by the PRR, and included a rail excursion to the iron works in Johnstown. ASCE's attendees and many of their wives were literally wined and dined, but no tours of the nearby ruined dam at South Fork were arranged. At Cresson, James Francis, whose name headed the author list in the report, told a reporter that Becker had stopped its release. Francis made his concerns known - he did not approve of holding back the report so long. A reporter wrote of the rumors and whispers he had overheard. "The opinion prevails that it is on account of his [Becker's] business associates in [Pittsburgh] that he takes his stand."

Six months later, at the start of 1891, another president took the helm of ASCE, Octave Chanute. He was a railroad man best known today as a pioneer of aviation, including his correspondence and association with the Wright Brothers in the early 1900's. Chanute's main income now came from consulting work on the railways. That could have been used by others to persuade him to let the report sit a while longer. The time was not yet ripe. On the other hand, William Shinn had given Max Becker custody of the sealed investigation report, so Chanute appears to have been "out of the loop" and likely could not have released it on his own authority. Finally, at ASCE's annual convention in Chattanooga in the summer of 1891, James Francis presented the long-awaited results. His fellow committee members did not attend. Worthen and Fteley, close friends, were likely peeved at the way their report had been delayed, edited, and handled. And no doubt Becker sought to avoid uncomfortable questions about why he had sat on the investigation for 2 years, when all that time hazards may have existed at other dams, jeopardizing the towns and people below them. The content of the report clearly shows it was sanitized and watered down, with statements that stretched the truth, favoring the railroad and the Club. It implied that the original dam built by the State had been fatally flawed, that the Club's modifications did not cause the failure. But in fact they did. The report also claimed that people who ran from the Day Express trains in East Conemaugh died because they did not stay aboard. The PRR's testimonials ordered by Pitcairn himself proved that wrong, as a group of passengers who got clear warnings about the dam ran and lived. Those who stayed in the cars and survived were lucky, many were not. This item of PRR liability was of no concern to Francis et al., but vital for Pitcairn, who must have inserted it through Becker.

Why would the hydraulic engineers, Francis, Worthen, and Fteley have gone along with all the content in the report, and the outrageous delay in releasing it? If they had tried to appeal to higher authority in 1890, that would have gone to the unsympathetic President Shinn. The railroad men could not pressure them directly. Becker and Pitcairn clearly relied on the polite discretion of these professional men, and their respect for the authority vested in the president of their Society. Francis and Worthen were past presidents, and Fteley was a serving Vice President. James Francis by his nature sought to avoid arguments in his work, but methodically went about his business. The illness that eventually overcame him had begun to place its mark on him, becoming very serious by 1892. He died late that summer.

And if posterity would ever grow wise about what the Club had arranged through Pitcairn, there were two convenient scapegoats. It would always appear that Max Becker, an engineer and German immigrant, was the one who sat on the report, and James Francis' name would appear as lead author for all the material content. But Becker was their pawn, willingly or not. James Francis had declared it was Becker's idea to hold the report until various court cases had been resolved, so ASCE would not become entangled in the legal system. Indeed, while the ASCE report was kept sealed for 2 years, renowned attorneys Philander Knox and his partner James Reed, who were Club members, easily fended off lawsuits in Pittsburgh, including one by a widow with eight children. I have no doubt that any appointed jury, Pittsburghers all, included railroad and steel workers who enjoyed a living wage, so long as a non-guilty verdict (or no verdict) would follow. As survivor Victor Heiser said many years later, "It is almost impossible to imagine how those [Club] people were feared." (McCullough 1968).

In the end the investigation report, as edited, was not unfavorable to the Club, so why delay its release? If I may say right here, the real truth of the matter was this - the railroad and Club men wanted to avoid at all cost having eminent hydraulic engineers like James B. Francis and William Worthen appear in court, place their hands on a bible, and swear to tell the truth as they saw it of the South Fork dam and the Club that owned it.

12.1 Scientific Conclusions

The amount of water in Lake Conemaugh at the moment the dam breached in 1889 was 1.455×10^7 m^3. The most likely range for the peak discharge was from 7200 to 8970 m^3 s^{-1}. Considering various sources of uncertainty, an upper limit for the peak discharge was 10,300 m^3 s^{-1}. The reservoir required more than an hour to drain, contrary to older claims that the lake drained in just 45 min. John Parke, the source of that drain time, wrote "I do not know the actual time it consumed in passing through the breach, but it was fully 45 minutes." His was a minimum estimate. If the entire complex breach formed rapidly, more than 65 min would have been needed to drain the lake to the floor of the upper breach. Part of the lake would still have been nearly 8 m deep at that time. Large areas of lake bed would have become exposed, and that explains why different observers reported a wide range in the time it took to drain the lake.

The dam-breach flood struck the South Fork rail station at 3:08 p.m. Given a time of dam failure close to 2:55 p.m. and arrival of the tsunami-like flood wave at Johnstown around 4:07 p.m., I estimate an *average* speed for the debris-filled flood wave of 0.32 to 0.33 km min^{-1} (~12 mph). The actual speed at any place along the flood's path depended on the valley shape, obstructions encountered, and the density of entrained debris. West of South Fork, the 80-foot arch beneath the viaduct bridge rapidly became obstructed with trees and other debris to form a short-lived dam and lake. Floodwater quickly backed up in the river loop upstream from the viaduct, surging against the arch supports. Unable to withstand the pressure, the viaduct collapsed, releasing the debris dam and unleashing a renewed flood wave.

Analysis of topographic data shows that the 1889 dam survey elevations by Francis et al. (1891) are systematically 1.9 m (6.2 ft) lower than the modern GPS reference frame. Another finding is that soil genesis on the bare rock spillway has produced as much as half a meter of soil in >120 years following the dam breach, indicating an average rate of soil accumulation >4 mm yr.$^{-1}$. However, the eastern spillway itself was accessible to vehicles until the property was acquired and protected by the National Park Service. Therefore, compaction by vehicles over time likely reduced the soil thickness that had built up on the spillway.

In repairing the dam, the South Fork Fishing & Hunting Club removed the five cast-iron discharge pipes beneath the dam. The pipes would have been pulled for John Reilly by the Club workers, as part of his deal to later sell the property to the Club. The Club also lowered the crest of the embankment, but not for the reported reason of widening the road on top. That "road" could easily have been widened without lowering the dam. The true reason for lowering the crest was to reduce the volume of material needed to fill the partial breach of 1862.

Unfortunately, lowering the crest of the dam rendered useless the auxiliary spillway on the western abutment and greatly reduced the discharge capacity of the main spillway. These actions cut the safe discharge capacity in half, from ~197 to only ~96 m^3 s^{-1}, fatally endangering the dam and the towns and people below it (Coleman et al. 2016). No qualified engineer would ever have approved the changes made to the dam.

Various witnesses reported that water levels in the river system had stabilized or even begun to drop by the early afternoon of the flood. The dam did not breach until just before 3 p.m., surviving more than 3 h of overtopping. Calculations show that if the dam had been maintained as originally built, overtopping would have been averted for many hours – long into the night… long enough for inflows to Lake Conemaugh to fall from their peak. The South Fork dam would never have overtopped and would readily have survived the 1889 storm and runoff event. The Johnstown Flood should never have happened in 1889.

Chapter 13
The Forgotten Dam: The Eastern Reservoir at Hollidaysburg and Its Role in Canal History

Neil M. Coleman and Stephanie Wojno

13.1 Introduction

The South Fork Fishing & Hunting Club and its former dam and lake will forever be notorious for their role in the devastating Johnstown flood of 1889. Nearly forgotten is the "sister" dam that impounded the Eastern Reservoir near Hollidaysburg, PA. This lake served the same function in supplying water to the Pennsylvania Main Line Canal system during periods of low river flows. The South Fork dam served the canal basin in Johnstown, west of the Allegheny Mountains, by adding flow to the Little Conemaugh River. The Eastern Reservoir in Blair County diverted water from the Frankstown Branch of the Juniata River. In the early 1800's that watercourse was known as the Southwest Branch of the Juniata. Feeder canals from the dam's spillway and sluice pipes added flow from that river to the Main Line Canal east of Hollidaysburg. Both the South Fork and Hollidaysburg dams were designed by the same engineer, William E. Morris, in the late 1830's.

We discuss the history of the eastern dam in relation to the Pennsylvania canal system. Our recent discovery of engineering notebooks in the Pennsylvania State Archive (PA Archive 2018) provided a wealth of information that chronicled the dam's construction. Due to careful preservation, most of the pages in those notebooks are still readable. They led us on a journey into the past to understand the site

N. M. Coleman (✉)
Department of Energy and Earth Resources,
University of Pittsburgh at Johnstown, Johnstown, PA, USA
e-mail: ncoleman@pitt.edu

S. Wojno
Johnstown, PA, USA

© Springer International Publishing AG, part of Springer Nature 2019
N. M. Coleman, *Johnstown's Flood of 1889*,
https://doi.org/10.1007/978-3-319-95216-1_13

preparation and building of the dam, spillway, culvert, feeder canals, and associated aqueducts over the river channels.[1]

LiDAR[2] data are now available for all of Pennsylvania. Vertical accuracy of LiDAR is discussed by the PA Dept. of Conservation and Natural Resources (DCNR 2018). We provide new images of the remains of the Eastern Reservoir dam and present a LiDAR analysis of the dam remnants, the areal extent, volume, and storage-elevation curve for the former lake, and the geometry and topography of its spillway. These scientific details have never before been published.

Due to safety concerns and a desire to recover the many tons of iron in its drainage pipes, the Eastern Dam was intentionally notched in 1882 at a time of low river flows, 7 years before the Johnstown Flood. We analyzed whether the draining of the lake was a wise decision. Could the Eastern Dam have survived the May, 1889 storm that destroyed the South Fork dam and Johnstown, killing more than 2200 people in the Conemaugh Valley?

13.2   History of the Eastern Reservoir

By the early 1800's there was a great need to improve commerce east and west across Pennsylvania, from Philadelphia to Pittsburgh and points west. The river systems allowed transport of goods east from Ohio to Johnstown. Numerous trading posts had been established along the watercourses. Construction of the Main Line Canal connected the east coast with the interior as far as Hollidaysburg through a combination of rail service, rivers, and canals. The State engineers completed work on the Juniata Division by 1832. According to the National Park Service (NPS 1991), the first canal boat to reach Hollidaysburg from Huntington was named the *John Blair*. It passed through the final set of locks and arrived to fanfare in Hollidaysburg on November 28, 1831. West of that town soared an imposing topographic barrier, the Allegheny Mountains, which terminated along their eastern flank at the steep Allegheny Front. Canals and locks could not ascend these mountains, so the 36-mile Allegheny Portage Railroad was built across Cambria County. Completed in 1834, it included 10 inclined planes, five to the east and five west of the mountains. The Portage Railway linked the Main Line Canal in Hollidaysburg with the canal basin and the Conemaugh River in Johnstown. This rather awkward transport system adequately supported trans-state commerce for a number of years. Persistence to overcome the challenges was rewarded by growth and the valuable brisk trade that ensued by connecting the towns and farms of the east and west.

---

[1] Our research relies on many nineteenth century publications and more recent work that document historic data using English units rather than SI units. For most calculations we use the always preferable SI units, but where we highly depend on old data sources we report the original English units. We believe this will help confirm our appropriate use of the nineteenth century data and will aid future workers who may further study the Eastern Dam.

[2] "LiDAR" is a portmanteau word combining "light" and "radar."

Hollidaysburg prospered at the canal terminus. Between 1831 and 1840 its population grew from 72 to 3000 (NPS 1991).

The canal basin at Hollidaysburg was originally planned to be built east of the town at the juncture of the Beaverdam Branch, the Frankstown Branch, and the small streams of Oldtown Run and Brush Run. This low-lying juncture is known as the Frankstown area (thought to be named for the owner of a trading post in the 1730's). At that location water supply would have been plentiful most of the time to support canal travel, because the Frankstown Branch of the Juniata flows through that valley. But the landowner would not agree to sell the Frankstown property that would have made an ideal canal basin. Engineers then considered a canal basin farther west on the south side of Hollidaysburg, fed by the Beaverdam Branch. Unfortunately, the smaller Beaverdam Branch by itself could not reliably furnish enough water for navigation, as its watershed covers only 87 square miles. An additional, reliable source of water was needed.

13.2.1   Site Surveys

In the early days of the canals the State engineers recognized that river flows often dropped so low during dry seasons that canal service was hindered. If good locations could be found, dams could be built to store water for release during dry times to keep up a reliable canal transport. In 1835 Sylvester Welch surveyed sites for both the Western and Eastern Dams. By July 15, 1838 sites had been approved by the Canal Board and oversight of the work was given to Principal Engineer William E. Morris.

Only the site above South Fork was seriously considered for the Western Dam. Three sites for the Eastern Dam were considered in the valley of the South (Frankstown) Branch of the Juniata River. Two of these sites considered but rejected were identified as "Sett's Mill" and "Lemars." Sett's Mill was the most distant and would have impounded the smallest lake. It was recognized that all the sites would have flooded good alluvial farmland, but a dam at Lemars would also have inundated a grist mill, sawmill, a tavern, a store with distillery, several dwellings, a covered bridge, and about one mile of turnpike road. The State would compensate property losses due to lake inundation, but undoubtedly at that time the prospective loss of a tavern and distillery would have been cause for general concern. Like Sett's Mill, the Lemars site was vetoed.

Ultimately, the easternmost site was chosen, with the proposed dam extending northward across the river valley from a property owned by Judge McEwen (Fig. 13.1). This location offered the largest impoundment by volume and would damage the least amount of property. McEwen's tract would also be closest to the Main Line Canal and basin in Hollidaysburg, thus requiring a much shorter feeder canal.

Figure 13.2 shows the location and areal extent of the Eastern Reservoir, 2.5 km southeast of Hollidaysburg, PA. The lake is now gone, but most of the embankment

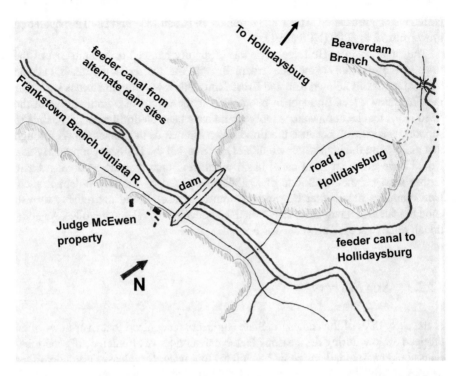

Fig. 13.1 Sketch of historic survey map of site chosen for Eastern Reservoir dam. The Main Line Canal was built at extreme upper right, on north side of channel. Original map in Pennsylvania State Archive (Morris 1839a)

still exists today on privately owned ground, with a prominent breach in its southern remnant through which the river flows. Reservoir Road parallels the southern shore of the former lake, preserving a memory of the once beautiful lake.

The proposed dam had two outlets for water, a "wasteway" (i.e., spillway) cut into bedrock at the northern end and five large drainage pipes that drained into a masonry culvert beneath the dam. The outflow then entered a feeder canal. In a presentation to the Canal Commissioners, William Morris (1839b) raised two issues about the McEwen site for the dam. First, the spillway had to be excavated on the northern end of the dam because the bedrock at the southern end had unacceptable quality. Second, the water surface in the planned feeder canal fed by the pipes would be 7 feet higher than the lake bottom, which meant the lowest part of the lake storage could never be used.

Another consequence of the site chosen was that the feeder canal had to be built north of the river in order to follow the natural contours around a large hill to the north. Only that way could water reach the Main Line Canal. An aqueduct would be required to convey water from the sluice pipes over the Frankstown Branch to reach the feeder canal. Costs for the McEwen site would include that aqueduct. And because the Main Line Canal was on the north side of the Beaverdam Branch, a second aqueduct would be needed to run the feeder over that watercourse.

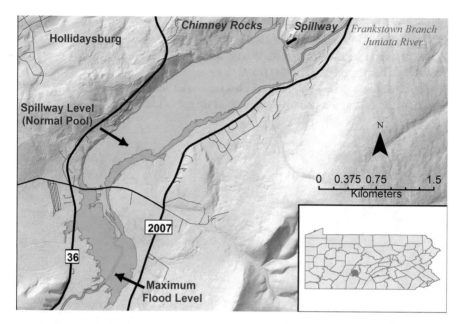

Fig. 13.2 Location map of the Eastern Reservoir. Area shown in light blue is "normal pool" or lake extent when its surface elevation equals crest of spillway floor at 959 ft. MSL. Darker blue is hypothetical lake stage and area of inundation for a flood level of 969 ft. (295 m) that would have overtopped the dam

13.3   Building the Dam

The dam was built between 1840 and 1847. State Engineer William Morris developed the plans for the Eastern Dam and its associated structures. Although construction was authorized by the Canal Commissioners to begin in February of 1836, work did not start then. Construction finally began in April, 1840 and as of October the work was progressing steadily (Board of Canal Commissioners 1841). Some of the cast-iron discharge pipes had already been made, and it was expected the rest would be ready before the spring of 1841. Due to scarcity of funds the clearing of ground at the base of the dam had not yet begun. "Clearing" meant removing and burning all organic debris on land that the dam would occupy, including trees, stumps, roots, even leaves, for a distance of 6 m beyond the dam base and spillways. About 10 acres had to be cleared for the base of the Eastern Dam. The cost estimated to build this reservoir was about $100,000 in 1840, not including office and engineering expenses and property damage that would result from eventually filling the lake and inundating private properties.

Due to severe funding shortfalls further construction was delayed for 5 years. The 1846 field book includes the name of the Superintendent of Construction, Engineer Thomas T. Wierman, Sr. He arrived at Hollidaysburg May 16, 1846 to begin work to finish the dam. Construction proceeded that summer and fall, and the contract for completing final work was placed on December 31, 1846. The work wrapped up by December of the following year. Final estimates of the work done by H. L. Patterson on the Eastern Reservoir and feeder canal were reported in 1848 (Fritz and Clemenson 1992, notes p 102).

Figure 13.3 gives Morris' original plan for the eastern dam, showing the design of the embankment, the control tower and drainage (sluice) pipes, and the location

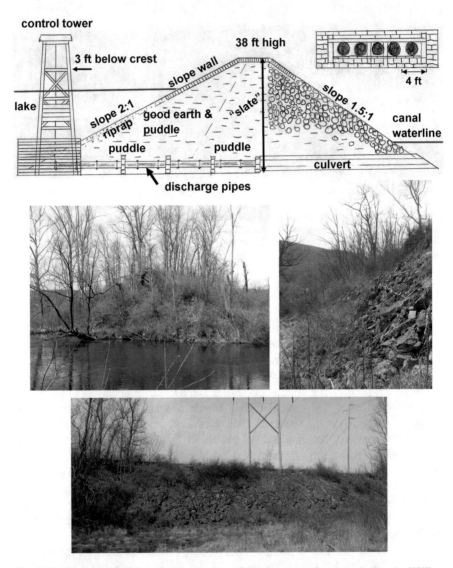

Fig. 13.3 Top: Plan for dam and control tower exhibited at time of contract letting, by William E. Morris (1846a). Center left: southern end (profile) of northern dam remnant at river margin. Center right: view looking south along downstream face of dam. Bottom: view of eastern side of dam (northern remnant) (photos 2016 by N. Coleman)

of the masonry culvert beneath the dam into which the pipes emptied. The maximum height of the dam was 38 ft. (11.6 m) from its base to the crest. The top of the dam was 10 ft. wide (5 ft. each side of center line) with paving stones. The downstream side of the embankment had a design slope of 1.5:1 and was covered with riprap consisting of "stones not less than 4 cubic feet each." The entire embankment still exists, except for the opening through which the Frankstown Branch flows. The GPS coordinates of the southern end of the embankment are 40°24′28.87″, 78°22′44.92″.

There were five horizontal drainage pipes made of cast iron, with inside diameters of 24 inches. Each of the 5 trains of pipes had 10 segments, each 7 ft. 7 in long. Accounting for joints each train was 70.4 ft. long, resting on masonry supports. The total amount of cast iron in the pipes was 60 to 70 tons. "Puddle" ditches under the pipes were cut 3 feet below the bottom of the pipes. The purpose of these ditches was to place clay layers of low permeability below and around the pipes to prevent water leakage from the reservoir side through the lower part of the dam. This was important to maintain long-term stability of the dam by preventing fine materials from washing out through the embankment.

The engineering notebooks (PA Archive 2018) record that the upstream ends of the pipes were laid half a foot inside the face of the wall. Pipe flow was regulated by rods fed through brass boxes and connected to stopcocks accessed from the control tower on the upstream side of the dam. In 1840 contractors were provided with a detailed diagram of a cast iron pipe, valve control rod, and brass box (State Engineer 1840). The gatekeeper used a boat to reach the tower to operate the stopcocks.[3] The pipes discharged into an arched masonry culvert 60 ft. long at the base of the dam that extended through the embankment to the downstream toe of the dam (see top of Fig. 13.3). The original plan called for stopcocks to also be placed on the lower ends of the pipes as well, which would have required the gatekeeper to enter and walk through the culvert to access the mechanism. However, the eventual waterline chosen for the feeder canal was higher than the culvert top, so access would not have been possible. The notebooks confirm that the pipe bottoms were 35.43 ft. below the top of the dam and the floor of the feeder canal was 30.43 ft. below the top of the dam. The pipe bottoms were 5 ft. below the bottom of the feeder canal and therefore they and the culvert would have been submerged.

In constructing the embankment the engineer used a sequence of 31 stations along a line from north to south, each station being 54 ft. apart. In this way the volumes of material needed and emplaced by contractors could readily be tracked and inspected. Station 22 was at the approximate center of the river channel at that time. At station 31 the top of the dam met the natural surface at the south abutment. The total length of the dam by these numbers would be 31 × 54 ft. = 1674 ft. or 510 m. William Morris had estimated that a dam at this site, 28 ft. high, would be 1230 ft. long on top. The dam as actually built was 38 ft. high and therefore longer on the crest.

---

[3] The distance over the lake from the embankment to the control tower was evidently too great to easily support a walkway of the type used at many dams after that time.

13.3.1   Damage During Construction

Significant washouts occurred during construction of the dam. These happened during river floods when the embankment at the river margin was eroded away. One such break occurred on October 13, 1846. A larger washout, referred to as a "big break" by the engineer in charge, happened on November 2, 1846. More than 4020 cubic yards of fill were needed to repair this break. The cast iron pipes and most or all the culvert had been installed by this time, but the Frankstown Branch had and still has normal flows that would greatly exceed the capacity of these five pipes to conduct all the water. Under low head conditions (1 m) they together could convey only 154 cfs (4.4 m^3 s^{-1}). Temporary spillways made of timber planking were needed to carry water safely over the new workings without eroding the bank as it was raised. However, even minor flooding could overwhelm the pipes and temporary spillways and would have remained a threat until the embankment was higher than the spillway floor. This is illustrated by present-day conditions in the watershed, which differ somewhat from the 1840's. For example, in 2017, the normally dry late summer produced a broad range of flows. Discharges of 90 to >400 cfs were seen at Williamsburg on the Frankstown Branch, 21 km downstream from Hollidaysburg. So even during the dry season, storm runoff could easily overwhelm the temporary spillways.

Additional embankment washed out in the winter of 1847. This too was overcome, and by the end of that year the engineer in charge and contractors had completed the work. Water immediately began flowing through a feeder to the canal basin in Hollidaysburg.

13.3.2   Spillway and Other Excavations

Initial investigations at the eastern dam had shown that the bedrock at the southern end was not suitable for a spillway. However, the northern end proved to be excellent for excavating a bedrock spillway. The high ground on the northern side of the valley was also excavated for dam material. Multiple quarry zones appear on the LiDAR map (Fig. 13.4), revealing that a hilltop was removed northeast of the spillway to obtain rock for the embankment and riprap cover. Some areas west of the embankment were excavated, which not only yielded dam material but also increased the depth and storage of the planned lake. The lower end of the spillway was dug out, which would have produced a small waterfall during discharges. Fig. 13.5 is a contour map of the spillway area.

Where it meets the spillway, the northern end of the long embankment curved slightly to the east. To prevent erosion, the floor of the spillway crest was excavated to bedrock and the northern end or flank of the spillway was also cut back to solid rock. The spillway was 150 ft. wide as designed. Where it joins the embankment, the southern end of the spillway appears to be solid rock along its lower flank. But

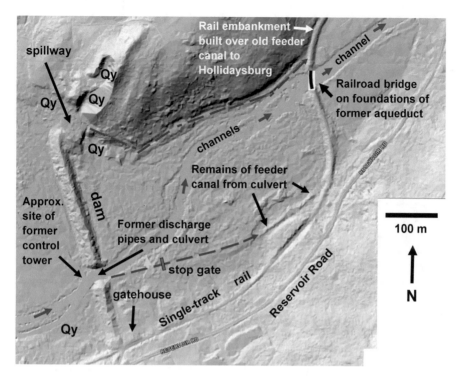

Fig. 13.4 LiDAR map of Eastern Dam. The symbol "Qy" shows locations of quarries on hillside and in former lake bed. Dashed blue lines are branches of feeder canal. "Old" northern feeder was excavated around base of hill, while southern feeder began at culvert and traversed floodplain east of the dam, crossing the river on an aqueduct. These two branch feeders merged north of the aqueduct. A railroad embankment now exists on the old canal right-of-way. Spillway on north side of feeder (*yellow arrow*) and stop gate below it (*blue bar*) are shown

higher up, the spillway "wall" appears to be part of the constructed embankment.[4] It is unclear what measures were taken to prevent hydrodynamic erosion of that part of the spillway wall, but it may have consisted of a slope wall, heavy riprap, or heavy planking material.

We found no documentation in the field notebooks about ancillary structures associated with this large spillway. Therefore some questions remain. For example, what kinds of wooden gates were used to divert spillway flow into the "old" feeder canal that ran along the base of the hillside to eventually reach Hollidaysburg (Fig. 13.4)? A plan does exist in the State Archive for a "waste weir in feeder," a structure 60 ft. wide with 8 gates, but its location and orientation are unclear. Its length was too short to have been placed directly across the spillway. Additional gates would have been needed to pass floods.

---

[4]Confirmation of this would require local excavations, which can only be done by or with the permission of the owner of this private property.

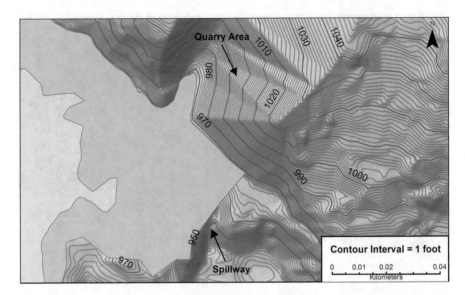

Fig. 13.5 Eastern Reservoir - LiDAR contour map of spillway and north end of embankment. Depression and terraces at top center and upper left were quarries mined to obtain embankment material and riprap. Light blue shows lake area with water surface at level of spillway floor elevation (959 ft). Darker blue shows added *lake* area at level that would overtop embankment (969 ft)

13.3.3   *Feeder Canals*

Vestiges of the two branch feeder canals exist today on the floodplain east of the dam and in the form of a wide path along the hillside from the spillway to the railroad bridge over the Frankstown Branch. These vestiges are represented by dashed lines in Fig. 13.4. The branch feeders merged to form a single canal just north of the aqueduct.

At the culvert mouth the discharging water flowed into a feeder canal that crossed the valley, curving northward to an aqueduct that carried the feedwater over the Juniata channel (Fig. 13.4). A cross-sectional view of the canal was preserved in a sketch in the original engineering notebooks (PA Archive 2018) showing its geometry as designed including its bottom width and side slopes (Fig. 13.6). A "puddle ditch" of clay layers helped to seal the bottom and side of the canal to reduce water loss from infiltration.

As recorded in the engineering field books (PA Archive 2018) the feeder canal began at the culvert mouth. An erroneous note in one notebook showed the feeder survey stations to be 27 ft. apart, but like those along the dam itself, the feeder canal stations were also 54 ft. apart. The line of the feeder was normal to the dam and extended straight for a distance of ~972 ft. The feeder began to slightly curve from stations 10 to 31, a distance of 702 ft. A strong curve in the course of the feeder then begins through station 40, covering 486 ft. From station 40 to 44 at the aqueduct the feeder is straight for 216 ft. These numbers are consistent with the curving path and

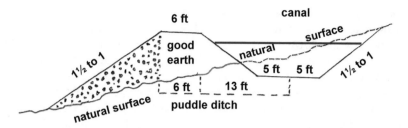

Fig. 13.6 Profile of feeder canal built east of dam leading from culvert to aqueduct over river. Canal bottom was 10 ft. wide. Puddle ditches at base of embankment were designed to reduce infiltration losses. Graphic based on sketch from engineering field book for Eastern Reservoir, circa June 1846 (PA State Archive 2018)

total length of the feeder canal as measured in our LiDAR map (Fig. 13.4). The trace of the feeder from the culvert shows distinctly in LiDAR up to its curving intersection with the present-day rail embankment, which was built on top of the old feeder canal all the way to Hollidaysburg. Topographic data show that the drop in elevation from the culvert waterline all the way to the Main Line Canal was >2 to 3 meters.

A second feeder canal began at the spillway and led eastward along the base of the hill on the north side of the valley. This spillway feeder is referred to in the engineering notebooks (PA Archive 2018) as the "old feeder." One notebook entry refers to the "canal from pipes to old feeder." We interpret the "old feeder" as the first one built, for the simple reason that as soon as water began flowing through the spillway it could be diverted into this branch canal and fed directly around the hill to Hollidaysburg. There a second aqueduct carried the feedwater over the channel of the Beaverdam Branch into the Main Line Canal. Feedwater could thus be sent to the Main Line even before the feeder from the culvert and its aqueduct over the Frankstown Branch were built. The spillway feeder could also supply canal water if any problems arose with the iron pipes and their stopcock mechanisms, so long as there was adequate flow in the spillway. We have not found records that show how water was diverted from the spillway into the north branch feeder. It must have incorporated a gate mechanism to allow for flood conditions, when much of the spillway flow would have to be diverted into the channel of the Frankstown Branch. The drop in elevation from the spillway floor along the "old" feeder to the Main Line Canal is about 9 meters.

We have reasoned that the Beaverdam Branch would usually have had little flow, because most of its flow would have been captured for the canal in Hollidaysburg. In the present day few parts of the original feeder canals remain to be seen.

One revelation from the engineering notebooks (PA Archive 2018) was the existence of a gate spillway in the side of the "canal from pipes to old feeder," near survey station #4. With stations 54 ft. apart, that would have placed the spillway 216 ft. from the culvert mouth. The spillway would have been placed on the north bank of the feeder canal so that the discharge would lead into the Juniata channel and under the aqueduct. The spillway gates would normally have been closed to allow flow from the pipes and culvert to proceed over the aqueduct and on to

Hollidaysburg. A smaller set of stop gates was installed across the canal, upstream from the aqueduct and 66 ft. below the spillway (Fig. 13.7). The engineering notebooks (PA Archive 2018) reveal these gates were 12 ft. below station #6, or 336 ft. from the culvert mouth. Closing these gates would have stopped flow to Hollidaysburg and directed all discharge from the pipes out through the canal spillway. Of course the pipes themselves could be opened and closed from the control tower. An engineering note says that riprap was placed on the feeder slope for 145 ft. from the dam to the spillway. The stones would have been placed on the north bank of the feeder to prevent erosion during times of spillway outflow.

13.3.4   Aqueduct

Flow from the pipes and culvert had to cross the Frankstown Branch of the Juniata River to reach the Main Line Canal in Hollidaysburg. An aqueduct was needed, as originally pointed out to the Canal Commissioners by Engineer William Morris. A sketch of the aqueduct foundations appears in a field notebook (PA Archive 2018) with a date on the prior page of August 6, 1846. Therefore, we believe the work on that structure occurred around that date. Fig. 13.8 shows William Morris' original plan for the aqueduct which included stone masonry abutments and a central pier, also of masonry. A railroad bridge now exists on the foundations of this old aqueduct, at 40°24'46.3", 78°22'28.1". A second aqueduct to the north, at approximately 40°25'43", 78°22'34", was needed to carry the feedwater over the Beaverdam Branch to the Main Line Canal.

A note in the engineering notebooks (PA Archive 2018) documents that the south pier of the aqueduct was excavated to 16.6 ft. below the bottom of the feeder canal. Engineers sought to place the foundations on solid ground. Therefore the excavation depth reveals the likely thickness of unconsolidated river deposits there in 1846. The aqueduct had a rectangular flow section 15 ft. wide with flared openings on both ends to accommodate the prismatic feeder canal, which had sloping sides and

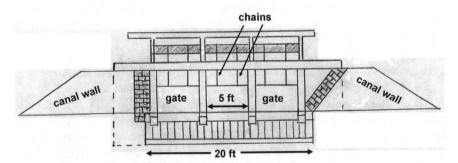

Fig. 13.7 Set of three stop gates placed across the feeder canal, between the culvert outlet and the aqueduct. Sketch based on diagram titled "Stop Gates in Feeder for Eastern Reservoir," by William Morris, exhibited at letting, 3/10/1846 (PA State Archive 2018)

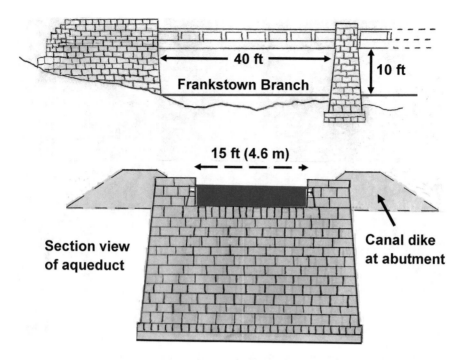

Fig. 13.8 Design of aqueduct at time of contract letting, March 10th, 1846. This structure carried feedwater from the sluice pipes and culvert over the Frankstown Branch of the Juniata River. A long feeder canal then conveyed the water to Hollidaysburg (after Morris 1846b)

a flat 10-ft wide bottom. A typical flow depth through the canal and aqueduct was ~3 ft. The design of the aqueduct may have included gates that could have been used to isolate the two branch feeders from each other. Those gates would have been helpful if maintenance work were needed on the aqueduct or the feeder from the culvert.

13.4   Operating the Eastern Dam

A gatekeeper who monitored and operated the dam was provided with a house. The house was east of the dam on the southern abutment. A present-day, private home exists on the site, and part of an original stone foundation appears to be preserved there (Fig. 13.9). A walkway along the top of the dam allowed easy access to the spillway. There would also have been a path north from the house to reach the feeder canal from the culvert. A walkway along the feeder gave access to the stop gates: one across the feeder itself and the other controlling flow through the feeder spillway on the north flank of the feeder canal.

Fig. 13.9 Present-day private residence at site of former Eastern Dam gatehouse (photo by N. Coleman)

Under normal conditions the feeder canal flowed 3 feet deep. When spillway flow was adequate part of that discharge could be directed into the northern branch feeder. At such times there was no need for discharge from the pipes and culvert. During dry times the spillway flow would not support a 3-foot flow depth in the feeder, and then the gatekeeper would ride in a boat to the control tower and open some or all of the discharge pipes. The water released would exit the culvert, flow east and north through the southern branch of the feeder to the aqueduct, then enter the long feeder canal to Hollidaysburg. If conditions were dry and lake inflows very low, the pipe discharges would have drawn the lake level below the floor of the spillway. Then the pipe discharge would be the sole source of water flowing through the feeder canal to Hollidaysburg.

During floods the flow to Hollidaysburg would be stopped, because no flow to the Main Line Canal was needed or desired. During severe floods, maximum outflow was needed from the dam to prevent overtopping and breaching of the embankment. All spillway flow was then directed into the river channel and away from the north branch of the feeder canal. The stop gates for the south branch feeder would be closed, the feeder spillway gates opened, and the pipes opened wide. In this way the combined flow from both the main spillway and the culvert could be directed into the Juniata channel, flowing under the aqueduct.

13.5 Decommissioning the Eastern Reservoir

By the 1850's new rail lines across Pennsylvania and more powerful locomotives allowed direct transport of goods and people across the state without the need for inclined planes and most canals. One of the new direct rail routes included the famous "Horseshoe Curve" near Altoona, an engineering marvel so strategically important it was on a German list of sabotage targets during World War II. Fortunately, "Operation Pastorius" failed.

In 1857 the Commonwealth of Pennsylvania sold the entire Main Line Canal system, including the South Fork dam and the Eastern Dam at Hollidaysburg, to the Pennsylvania Railroad Company. After the railroad achieved direct service across Pennsylvania, the canal system continued to be used for local service but was no longer the primary mode of transporting goods across the state. The canals, locks, and dams were not needed by the railroad and thereafter were neglected with minimal oversight. No funding was available to repair recurring flood damage to the works.

Local citizens had safety concerns about the Eastern Dam, so it was periodically drained. The *Johnstown Daily Tribune* (1880) printed an article about happenings at the dam, reporting that the "old reservoir" near Hollidaysburg had been drained off in April that year. Some five wagon loads of fish were pulled from the mud of the lake bed. Local people had begun to fear the dam might break someday, flooding a large area of the country. Mr. Wierman of Harrisburg, the engineer who supervised most of its construction, had been consulted about its safety. He said the dam would have lasted at least 100 years to come.

But the concerns of the local citizens held sway. Two years later the dam was "notched" at a time of low water levels to keep the lake from reforming. The *Johnstown Daily Tribune* (1882) reported this in their edition of Feb. 7:

> The task of destroying the reservoir at Hollidaysburg was completed last week. The reservoir was built by the State for supplying the canal in low water. It was a beautiful sheet of water, two and one-fourth miles long, covering nearly 700 acres, had stood for 50 years and its breast wall was strong enough to stand a 100 years longer, and it was a pity to destroy it. Mr. Gayton, the P.R.R.'s [Pennsylvania Railroad's] agent for doing the work, dug a channel through the breast of the dam down to the water's edge in such manner that the water will cut a channel to the bottom in 2 years or less and then the beautiful sheet of water will disappear forever. It took 5 weeks to dig the channel to the water's edge.

Fritz & Clemensen (Fritz and Clemenson 1992 p.72) reported that the Eastern Dam had been notched on February 10, 1882, but as revealed by the *Tribune* article this work began in early January. The notch was very effective in preventing a deep lake from forming and allowed subsequent flows to slowly erode and safely widen the breach in the dam. Perhaps locals knew about the partial breach of the South Fork dam years earlier, in 1862, and understandably had grown leery of their local dam. It is curious that the Eastern Dam was drained around the same time the South Fork Fishing & Hunting Club was repairing their newly acquired dam in the hills above Johnstown.

The notch was created where the embankment was highest, near the control tower, so that erosion of the embankment there would eventually expose and allow easy recovery of the cast iron discharge pipes. We found no evidence that the pipes and culvert remain in place at the base of the dam, because the wide river channel now occupies their former location. For that reason we have not done a magnetometer survey to confirm their absence. The 60+ tons of iron in the pipes and other hardware would have had considerable value at that time after the dam itself was no longer needed. We have not determined who recovered the iron pipes, but the Pennsylvania Railroad would have had the resources to easily do so. The sale of those many tons of scrap iron would have more than repaid the company for the labor to notch the dam and recover the iron.

13.6   Lake Volume During Operational Period

We used LiDAR data that are now available for all of Pennsylvania to show high-resolution details of the topography of the dam, spillway, and the former lake basin. The fact that the lake is no longer present makes it easy to accurately measure the former size and volume of impounded water in the lake, based on LiDAR elevations of the spillway floor and the dam crest. Fig. 13.10 shows the storage elevation curve for the Eastern Reservoir. The curve reflects the nonlinear increase in volume stored as the lake level rises to an overtopping level.

The storage elevation curve was used to evaluate engineer Morris' assertion that only a small amount of the lake volume would be unusable to feed the canal, since the feeder canal had to be built up high enough to cross the river channel through an aqueduct. Morris was correct. Only 2.3% of the lake volume was below the level of the pipe intakes when the lake was at the level of the spillway floor (normal pool).

Data Summary	
Hydraulic Data for the Eastern Dam and Reservoir	
Maximum height of embankment (measured in 1846)	38 ft. (11.6 m)
Length of embankment	1674 ft. (510 m)
Present elevation of northern remnant dam crest (from LiDAR)	969 ft. (295.4 m)
Height difference from top of dam to bottom of sluice pipes	35.4 ft
Height difference from bottom of feeder canal to bottom of pipes	5.0 ft
Height difference from bottom of feeder canal to base of dam	4.56 ft
Height difference from top of dam to bottom of feeder canal	30.43 ft
Spillway floor elevation (from LiDAR)	959 ft. (292.3 m)
Surface area of lake –	
At normal pool elev. of 292.3 m	19.712×10^6 ft^2
Near overtopping of crest	29.093×10^6 ft^2
Storage volume of reservoir –	
At normal pool elev. of 292.3 m	6.088×10^6 m^3

(continued)

Data Summary	
Near overtopping of crest (elev. 969 ft.; 295 m)	12.839×10^6 m³
Combined discharge rate for 5 sluice pipes (lake level at dam crest) (max. head = 35 ft.; pipe bottoms were 35.5 ft. below crest)	>500 cfs (14.2 m³ s⁻¹)
Discharge rate for 5 sluice pipes at low water level (3.3 ft.; 1 m)	147 cfs (4.2 m³ s⁻¹)
Width of spillway at northern end of dam	150 ft
Maximum spillway flow capacity at overtopping lake stage	327 m³ s⁻¹
Maximum safe discharge capacity of spillway and pipes	~341 m³ s⁻¹
Estimated peak discharge rate of flood if dam had failed at overtopping	~20,000 m³ s⁻¹

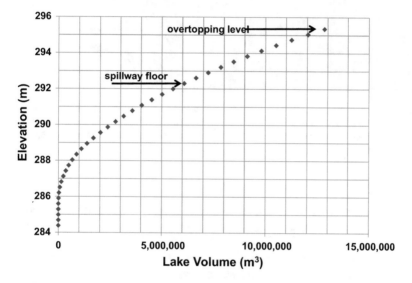

Fig. 13.10 LiDAR-based storage-elevation curve for the Eastern Reservoir near Hollidaysburg. Highest point on the curve represents an overtopping level of 969 ft. Arrows point to lake volumes impounded when the lake surface equaled the spillway floor elevation (normal pool), and also the elevation at which the lake would overtop the dam. LiDAR data source: PASDA (2013a, b)

13.7   Hydraulic Analysis of the Spillway and Discharge Pipes

The dam was intentionally notched in 1882 at a time when the river discharges would have been relatively low. Based on what we know today, was this intentional breach of the Eastern Dam a wise decision? Could heavy rainfalls and snowmelt have threatened to breach the dam in the 135 years since the dam was drained in 1882? We analyzed the combined discharge capacity for the spillway and sluice pipes and compared it to the largest historical flood on record. We evaluated the spillway geometry using LiDAR data (Fig. 13.11).

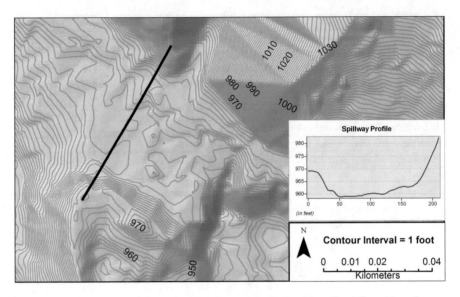

Fig. 13.11 LiDAR topographic profile of spillway at the Eastern Dam. Some deposition of material has taken place on the floor of the spillway. Original design called for a spillway 150 ft. wide and 10 ft. deep. Top width of the spillway was 170 ft., bottom width 130 ft., and the width at a depth of 5 ft. was 150 ft

The Eastern Dam had a far larger drainage basin than that of the South Fork dam, about 200 vs. 50 square miles, but it had the same number and size of discharge pipes and a spillway 150 ft. across. The bigger drainage basin meant that the Eastern Dam could receive much larger river discharges from runoff than the South Fork dam, given similar rainfall depths over the basins.

Equation 11b of Walder and O'Connor (1997) was used to evaluate maximum flow through the spillway at the Eastern Dam. LiDAR data reveal that the present spillway dimensions have been altered by deposition of colluvium. The northern half of the spillway is now only about 7 ft. deep. The original spillway dimensions were 130 ft. across the bottom, mean width of 150 ft., depth of 10 ft., and side slopes of ~25 degrees. Given these values, the flow capacity of the Eastern Dam spillway was ~327 m^3 s^{-1}. The calculation assumes the full width and depth of the spillway was unobstructed. To obtain the safe discharge capacity of the dam, the capacity of the discharge pipes must be added.

We estimated the maximum flow through the discharge pipes using an analytical equation for the inflow capacity of a pipe spillway (Chow 1964), where

$$q = a \sqrt{\frac{2gH}{1 + K_e + K_b + K_p L}} \qquad (13.1)$$

q = discharge rate (cfs) (1 m^3 s^{-1} = 35.3 cfs)

   a = cross-sectional area of pipe (ft^2) (3.14 ft^2)

   g = surface gravity (32.2 ft. s^{-2})

   H = total head (ft)

   K_e = coefficient for entrance loss [0.5 when pipes are mounted flush and do not protrude into upstream water column]

   K_b = coefficient for losses due to bends (= 0 for straight pipes)

   K_p = coefficient for pipe friction loss

   L = length of pipe (ft).

Given a cross-sectional flow area of 3.14 ft^2, a coefficient for pipe friction loss of ~0.01 (appropriate for the cast iron used in these pipes), an entrance loss coefficient of 0.50, a total head of ~31 ft. (with the dam near overtopping, pipe intakes ~35 ft. below the dam crest, and reducing the head 4 ft. for culvert submergence effects), and a horizontal pipe length of 70.4 ft. (21.4 m), each pipe could discharge up to ~2.68 m^3 s^{-1}. Together the five pipes could have transmitted up to 472 cfs, or ~13.4 m^3 s^{-1}. This would have been the maximum flow rate for unobstructed pipes. Some flow loss may have occurred depending on the geometry of the flow control device (stop-cocks) at the upstream side of the pipes.

The total safe discharge capacity of the dam was 327 m^3 s^{-1} from the spillway +13 m^3 s^{-1} from pipes = ~340 m^3 s^{-1}. Historic floods on the Frankstown Branch of the Juniata River have been documented for Williamsburg, downstream from Hollidaysburg. A USGS (2017a) stream gage exists there, and the drainage basin above this gage is 293 square miles. The drainage area above the Eastern Dam is less than this, about 200 square miles. The largest historic flood at Williamsburg was on June 1, 1889, with an estimated peak discharge of 35,500 cfs, or ~1005 m^3 s^{-1} (USGS 2017b). The second largest historic flood was on March 18, 1936, when the discharge reached 30,000 cfs. As a first-order estimate, the 1889 flood inflow to the eastern reservoir, based on drainage area fraction, would have been at least 685 m^3 s^{-1} and perhaps more, since the drainage area above the Eastern Dam has a higher relative proportion of mountainous terrain which promotes rapid runoff. This estimated inflow is *twice* the safe discharge capacity of the Eastern Reservoir, so unless the lake level had been exceptionally low before the storm (i.e., had a large buffer volume to fill before overtopping) we find that the dam would have experienced overtopping for a significant time that would have eroded the downstream flank of the dam and most likely would have breached it. The reservoir would not have been low because May 1889 was not unusually dry in western and central Pennsylvania. For that month, Erie and Pittsburgh had reported 14 days with rainfall and Harrisburg 12 days (Dunwoody 1889).

We conclude that the Eastern Dam would have suffered the same fate as the South Fork dam, during the same spring storm event in May of 1889. It was therefore a wise decision to notch the Eastern Dam in 1882, draining the lake 7 years before the Johnstown flood and averting a catastrophic dam breach flood.

13.8   Hypothetical Dam Breach

It is further possible to estimate the flood magnitude of a hypothetical breach in the Eastern Dam, at a time when the reservoir would have been overtopped and breached by a major flood like the one on June 1, 1889. Several analytical methods exist to do so. One technique uses empirical data from the breach of many natural and human-made earthen embankments around the world for which peak discharges have been estimated. If these are plotted on a logarithmic chart, relating the volumes of the lakes vs. the estimated peak discharges, the result is a regression equation and prediction line that broadly estimates the peak flow for any earthen dam breach based on its lake volume (Fig. 13.12).

Using a volume of 12.839×10^6 m^3 for the Eastern Reservoir at overtopping, we estimate a peak discharge of 1000 to >5000 m^3 s^{-1} for a hypothetical breach of the Eastern Dam. A large fraction of the flood wave would have spread out over the floodplain east of the dam and proceeded down the main river valley. Backwater from the hypothetical flood would have reached Hollidaysburg and caused substantial inundation, although the town would have had some protection by not being in the direct path of the main river valley below the dam. We have not modeled the

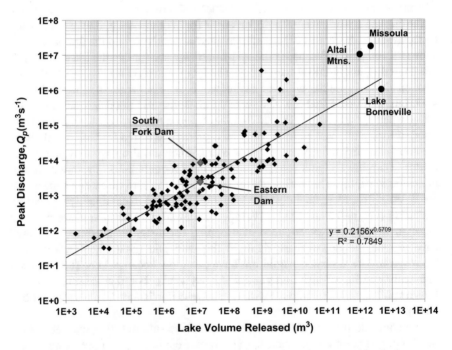

Fig. 13.12 Plot of peak discharge vs. lake volumes for more than 120 terrestrial floods from natural and rock material dams (O'Connor and Beebee 2009). Data points added for Johnstown flood, hypothetical breach of Eastern Dam, and three terrestrial megafloods. Fitted line is a power regression equation

maximum flow depth that could have occurred in Hollidaysburg. Severe damage would have occurred downstream in low-lying areas of Williamsburg, PA and probably farther downriver at Alexandria.

13.9   Conclusions

The Main Line Canal system played a key role in supporting commerce across central Pennsylvania in the years 1834 to 1854. In the Juniata Division it was recognized that low river flow during dry seasons did not adequately support boat transport. The Eastern Dam was completed in 1847 to augment flow to the canal basin in Hollidaysburg. The dam was designed by engineer William E. Morris and its construction supervised by several other engineers. Field notebooks maintained by the Pennsylvania State Archive (2018) in Harrisburg preserve the historic record of the building of the Eastern Dam. Our scientific analysis of the characteristics of this lake and dam sheds new light on the technology and history of the role of this reservoir in the Juniata Division of the canal system. The reservoir at normal pool level (base of the spillway floor) impounded 6.088×10^6 m^3, and if it had reached overtopping levels would have impounded 12.839×10^6 m^3. We developed a storage-elevation curve to show the impoundment volumes at various lake levels.

By 1854 the Eastern Dam and its counterpart, the "Western" or South Fork Dam, were becoming obsolete due to new rail lines and rapidly advancing steam engine technology. The famous "Horseshoe Curve" west of Altoona was completed that year. The complex system of the Portage Railroad, with its inclined planes, would soon no longer be needed.

Seven years before the Johnstown flood, in January and February of 1882, the eastern dam was "notched" during a time of low river flow. That action prevented the lake from reforming and almost certainly averted a future disaster. The flooding on the Frankstown Branch of the Juniata that occurred on June 1, 1889 would have overtopped the Eastern Dam for a significant time and most likely would have breached it. The decision to "notch" the dam in 1882 was a wise one, and also permitted recovery of 60+ tons of cast iron in the former discharge pipes beneath the dam.

Most of the embankment of the Eastern Dam remains today on privately owned land. A right-of-way for electrical transmission lines passes east to west over the dam. The spillway is also intact, and the foundation for the former aqueduct over the Frankstown Branch remains, supporting a railroad bridge. A single-track railway crosses this bridge and the floodplain east of the dam. It follows the former feeder canal which was filled in and covered by the rail embankment.

The remnants of the Eastern Reservoir still stand as monuments to the skill of engineers in the mid-1800's. They are historic relics of the early transportation system across Pennsylvania that fostered commerce and trade, until superseded by the trans-state railways.

References

Board of Canal Commissioners (1841) Annual Report of the Board of Canal Commissioners for the year ending 31 Oct. 1840, read in the state Senate on Jan 15, 1841, p 46

Chow VT (ed) (1964) Handbook of applied hydrology. McGraw-Hill Book Co., NY, 1418 p

DCNR (2018) PA Department of Conservation and Natural Resources, link to vertical accuracy reports for LiDAR data in Pennsylvania. Available at: http://www.dcnr.state.pa.us/topogeo/pamap/documents/index.htm. Accessed 5April 2018

Dunwoody HHC (1889) Monthly Weather Review, vol XVII, No 4, p 118, Washington City, May 1889

Fritz D, Clemenson AB (1992) Special study – Juniata and Western Divisions, Pennsylvania Main Line Canal. For National Park Service, US Department of Interior, 132 p

Johnstown Daily Tribune (1880) Story under "Local Items – Minor." Friday April 16, 1880. Vol. VIII, No. 39, p 4

Johnstown Daily Tribune (1882) Story in edition of Feb 7, 1882

Morris WE (1839a) "A Map of S. Branch of Juniata River Surveyed with a view to Location of Eastern Reservoir," by Wm. E. Morris, Civil Engineer, November 1839. Pennsylvania Canal Maps, map book 18, page k, 2 sections, available at: http://www.phmc.state.pa.us/bah/dam/rg/di/r017_0452_CanalMapBooks/CanalMapBook18Interface.html [Shows: Furnace, Saw Mill, Sett's Mill, Distillery, Turnpike, Newry, Dam, Feeder, Penna. Canal, Basin, Hollidaysburg]. Accessed 26 March 18

Morris WE (1839b) Report of William E. Morris, Engineer [No. 12]. Reservoirs. Engineer's Office, Hollidaysburg, Nov. 1, 1839. To: James Clarke, Esq., President of the Canal Board, p 125–131

Morris WE (1846a) PA State Archives, PA Canal Maps, map book #18, page w, map dated Mar 10 1846, available at: http://www.phmc.state.pa.us/bah/dam/rg/di/r017_0452_CanalMapBooks/CanalMapBook18Interface.html. Accessed 26 March 18

Morris WE (1846b) PA State Archives, PA Canal Maps, map book #4, page 11, plan dated Mar 10 1846, available at: http://www.phmc.state.pa.us/bah/dam/rg/di/r017_0452_CanalMapBooks/CanalMapBook4Interface.html. Accessed 27 March 18

NPS (1991) Reconnaissance survey – Juniata River Corridor, America's Industrial Heritage Project, Southwestern Pennsylvania. Southwestern Pennsylvania Heritage Preservation Commission. Prepared by National Park Service, Sept. 1991, 86 p

O'Connor JE, Beebee RA (2009) Floods from natural rock-material dams. In: Megaflooding on earth and Mars. Cambridge University Press, New York, pp 128–171

PA Archive (2018) Three field books for the Eastern Reservoir, one by Superintendent of Construction, Engineer Thomas T. Wierman, Sr., 1846. W2-b. 40. Records of the Board of Canal Commissioners, General Records, Engineering Records, Field Book (Eastern Reservoir), microfilm positive roll #16, RG-17

PASDA (Pennsylvania Spatial Data Access) (2013a) Download site for PAMAP LiDAR: http://www.pasda.psu.edu/uci/SearchResults.aspx?originator=%20&Keyword=pamap%20lidar&searchType=keyword&entry=PASDA&condition=AND&sessionID=5775552642015710143015. Accessed 4 April 2018

PASDA (Pennsylvania Spatial Data Access) (2013b) Metadata summary for PAMAP program 3.2 ft digital elev. model of Pennsylvania (based on LiDAR data). Available at: http://www.pasda.psu.edu/uci/DataSummary.aspx?dataset=1247. Accessed 4 April 2018

State Engineer (1840) Plan of cast iron pipes (and other hardware) for Eastern Reservoir furnished to contractors 4/24/1840. Available at: http://www.phmc.state.pa.us/bah/dam/rg/di/r017_0452_CanalMapBooks/CanalMapBook4Interface.html. Accessed 27 March 2018

USGS (2017a) Online data for Station 01556000, Frankstown Branch of Juniata River, at Williamsburg (drainage area 291 sq. miles), river discharge, stage, and precipitation data available at: https://waterdata.usgs.gov/usa/nwis/uv?01556000. Accessed 24 March 2018

USGS (2017b) Historical peak flows for Station 01556000, Frankstown Branch, Juniata River, at Williamsburg, from: https://www.weather.gov/media/marfc/FloodClimo/JUN/Williamsburg.pdf. Accessed 24 March 2018

Walder JS, O'Connor JE (1997) Methods for predicting peak discharge of floods caused by failure of natural and constructed earthen dams. Water Res Res, v 33, no 10, 2337–2348

Chapter 14
End Notes

14.1 The Tale of Leroy Temple

The official death toll of the 1889 flood has long been published as 2209, but this number must be reduced by one. The *Johnstown Daily Tribune* on Saturday, August 18, 1900 (which cost 2¢) wrote about a survivor who had been reported as "missing and assumed dead." That week, Mr. Leroy C. Temple showed up alive in Johnstown. He had returned to visit old friends and to let them know he was not dead and was still "a lively Yankee."

On the day of the flood Mr. Temple, affectionately known to his local friends as "Tod," was caught in the torrent and swept downstream, to be trapped in the massive debris jam at the stone bridge over the Conemaugh River below Johnstown. McCullough (1968) suggested that Leroy pulled himself up, saw what remained of Johnstown, then walked out of the valley. He did leave, but not right away. Leroy Temple was seriously injured but managed to crawl out of the debris. He had been pinned near the bridge and had enough strength left to climb out using the suspended rails for support after the embankment on the east side of the stone bridge had washed out. Within 3 weeks rail communication had been reestablished, and Leroy had recovered enough to travel. He and his family then left for Massachusetts, his native state.

Since the time of the flood Mr. Temple had been living in Beverly, Massachusetts, employed there by various companies. We now know a lot about this man who "returned from the dead." In 1893–94 he was an "ice teamster" living on #99 Balch Street. There were three different ice companies in the area at that time supplying blocks of ice for use in early refrigerators and cold storage rooms. During 1895–1898 Temple was listed as a "driver," living the first 2 years in a house on #16 Cliff Street and later on #4 Cliff Street.

The 1900 census shows Temple and his family living at #11 School Street in downtown Beverly, near the intersection with Rantoul Street. Leroy's first name in that census is given as "Roy," born September 1857, and his occupation is "teamster."

© Springer International Publishing AG, part of Springer Nature 2019

N. M. Coleman, *Johnstown's Flood of 1889*,

https://doi.org/10.1007/978-3-319-95216-1_14

Both Leroy and his wife Viola M. Temple were 43 years old. Viola was a nurse with four children from a prior marriage named Harry, Alice, Edith, and Wilhelmina Schade. Wilhelmina was the youngest at 15, and all four of Leroy's step-children had jobs related to the nearby shoe factory.

In 1903–04 Temple was a "driver" living on #14 School Street in Beverly. In 1905 he was still a driver but had moved to # 5 Phillips court. By 1906 he had his own company "L.C TEMPLE Piano and Furniture Moving." Temple ran this company from various addresses until 1912. The following add ran that year in a local newspaper:

L. C. TEMPLE

FURNITURE AND PIANO MOVING

EXPRESSING OF ALL KINDS

Orders Promptly Attended To

Leroy C. Temple, Proprietor, Rear 27 Park St., Beverly

Residence, 58½ Pleasant St.

By 1913 Leroy Temple was reported "deceased" in the Beverly City Directory. After he passed, his widow remained for a time in Beverly and in 1915 was rooming at #40 Pond Street. The following year Viola Temple moved from Beverly to Bellefonte in Pennsylvania, the state of her birth.

I knew about this story and had always imagined that Leroy Temple returned in 1900 because he saw pictures from the devastation that year of Galveston, Texas. Those images of demolished buildings and piles of debris would have looked like the scenes of destruction he witnessed in Johnstown. But Mr. Temple came back to Johnstown in mid-August of that year, weeks before the Category 4 hurricane struck Galveston on September 8th, killing at least 6000 people and perhaps thousands more. Leroy might have learned that his name was on a list of 1889 flood victims and come back to clear that up and see how his old friends from Johnstown and their families had fared in the 11 years since the flood.

It is good to read of the happy ending for Leroy Temple and his family. I wish there were others who were assumed dead but could have come back long after the flood. Perhaps a few survived who chose that moment to disappear and start new lives in new places with the same or a different name. At that time there was much romantic appeal in traveling to the western territories to begin anew. The Oklahoma Land Rush began on April 22, 1889, just 5 weeks before the flood, and there was still great interest in striking new deposits of gold and silver in the West, in Canada, and elsewhere. The West was the place to go if life took a turn for the worse. It was a destination for the lonely and broken-hearted, for those who lost their jobs, and for those who dreamed of riches. Many doctors advised tuberculous patients, "consumptives," to escape the damp mountains of the east and seek the western lands if they were able.

One of my own relatives, Lieutenant George William Hay, survived a gas attack in France in WWI that injured his lungs. After returning to Pennsylvania he eventually settled in New Mexico for health reasons. He had a law practice in Silver City and in 1930 became a district judge in Grant County, riding on horseback from town to town carrying a sidearm for protection. The West was a seventh heaven where anything might be possible. And in 1889 life had indeed taken that turn for the worse for many survivors of the Johnstown flood.

For a time articles appeared about family and friends reaching out for any word of missing loved ones. On July 5th the *Johnstown Daily Tribune* printed a note from G. M. Green of Blairsville asking for any information about Mrs. Abraham Wilkinson, who was supposed to be living in Johnstown. And more than a year after the flood, the *Tribune* (June 25, 1890) reprinted the following from the Greensburg *Argus*:

> We have a brother and we have not heard from him directly for over a year, but lately we have heard he was in Johnstown the day before the flood....This brother used to be in Westmoreland County and on the Monongahela River. He was a cripple and drove around with a dog and wagon. Any one that can give any information that will help us to learn whether he got out of Johnstown, or his present whereabouts, will confer a favor on his father and brother, Thomas and Robert Forsyth, New Castle, Lawrence County, Pa....He was called Frank Forsyth, and his name was painted on the wagon. He was a peddler and repairer of clocks and watches.

The names of Mrs. Wilkinson and Frank Forsyth never appeared in the lists of flood victims. Perhaps they survived but I find no record of them. No one really knows how many died in the 1889 flood as there were travelers like Forsyth passing for business or pleasure, and also a small transient population including itinerants with no local family and of whom there was no record. Hundreds of recovered bodies were so degraded that they could not be identified, and between 60 and 70 victims, possibly more, were never found. Perhaps evidence of another survivor will appear someday, in a dusty diary in an old trunk in an attic corner that will tell the tale. Or maybe an old record will emerge of a death-bed confession from a survivor who had changed their name and moved on.

But Leroy Temple did survive, and was "a lively Yankee yet," so the official death toll from 1889 must be reduced to 2208.

14.2   Clara Barton and Nathaniel Deane

Her full name was Clarissa Harlowe Barton. The steel of her nature was hardened in the fires of Civil War battles, where tens of thousands of wounded soldiers lay in the fields or struggled to reach aid. The medical staffs of both the northern and southern armies were overwhelmed with the vast numbers of casualties. In stepped Clara Barton, who believed she could do more than collect medical supplies at home; she went in person to the places of battle, to aid the wounded, sometimes at risk to her own life. She arrived at the bloodbath of Antietam with several wagonloads of desperately needed medical supplies, and while there a bullet passed through her garment. With her knowledge of disasters and mass casualties, Barton founded the American Red Cross in 1881.

Six months before the Johnstown Flood, Clara Barton had traveled to Glen St. Mary, Florida, to honor the heroic work of a small band of Red Cross nurses. Several months before her trip, ten nurses arrived in MacClenny, Florida, near Jacksonville, to valiantly aid the victims of a terrible outbreak of yellow fever. The train would not stop near the town due to fear of the epidemic,[1] but slowed a mile past it and the nurses, seven women and three men, jumped from the car. It was night and pouring rain, but holding hands in the dark they made their way along the wet tracks to the town. They found all the physicians there were ill, and set about their work caring for the sick. A physician from New Orleans, Dr. Gill, eventually arrived to lead their efforts, and their aid soon extended to other towns, including Sanderson, Glen St. Mary, and Enterprise (Barton 1910).

Barton herself arrived on November 24, 1888. She wrote of the exploits of the "MacClenny Nurses" and that the Red Cross was "...glad of the two or three [subsequent] months in which no call for action was made upon us..." But then the following spring came the terrible news from Johnstown. She later wrote "...so frightful and improbable were the reports, that it required twenty-four hours to satisfy ourselves that it was not a canard." (Barton 1910) She rapidly mobilized every resource of the Red Cross.

Thanks to the efforts of Robert Pitcairn and the Pennsylvania Railroad Company, rail service was restored from Pittsburgh to Altoona by June 13th. That remarkable work included the repair of many washouts, 3 miles of destroyed tracks and embankments, and the erection of large wooden trestles to temporarily replace the destroyed Bridge #6 and the stone viaduct at the river loop west of South Fork. Clara Barton and a large team of doctors and nurses of the Red Cross had arrived in Johnstown on Wednesday, just 5 days after the flood. She soon met Daniel H. Hastings,[2] Adjutant General of Pennsylvania, who had taken charge of recovery and relief efforts. Barton announced that the Red Cross had "arrived on the field." But Hastings did not know what that meant or the caliber of this woman, who was short only in stature. His immediate thought was to make comfortable this poor, lone woman who must naturally be helpless in the face of the disaster around them. As she later wrote in her 1910 book (p 159),

> It was with considerable difficulty that [Hastings] could be convinced that the Red Cross had a way of taking care of itself at least, and was not likely to suffer from neglect. I don't believe he quite got over his mistrust until a week later, when carloads of lumber from Iowa and Illinois began to come in consigned to the president of the Red Cross. As this was the only lumber that had come, the military were constrained to "borrow" from us in order to erect quarters in which to entertain the Governor of the State on the occasion of his first visit.

---

[1] The true cause of yellow fever was not yet known, but Clara Barton wrote a keen insight, that "frost put an end" to the epidemic. Cold weather knocked down the clouds of mosquitoes that bore the disease.

[2] Hastings became Adjutant General of the Pennsylvania militia in 1887, with the rank of Major General. He came to public attention for leadership in Johnstown's recovery. He ran for Governor twice, winning in 1894.

The work of the Red Cross in Johnstown has been noted by various authors but is best described in Clara Barton's own words. Her 1910 narrative leaves the reader with a great appreciation of the early work of the Red Cross and its legacy. I do not find it self-aggrandizing. She thanks and credits many people and organizations for their work in a way that would inspire, then and now, others to volunteer in times of need. One of the many persons she met in Johnstown was Nathaniel Carter Deane. She long remembered working with him there. At a future army regimental reunion she called him "Comrade Deane, our boy I met at Johnstown." (American Historical Society 1922)

Nathaniel Deane had risen from a railroad laborer to become a terminal manager near Pittsburgh with the Pennsylvania Railroad. In 1917 he was living in Carnegie, PA, retired from the "Company" where he had worked for 52 years. Although they were both involved in the relief work after the 1889 flood, Deane and Clara Barton had an earlier connection as well. Deane was a Civil War veteran from Massachusetts, having served with the 21st Regiment of the Massachusetts Volunteer Infantry. Twice wounded in his first battle at Roanoke Island, he fought in many battles, major and minor, and was wounded again at Camden, at the siege of Knoxville, and twice more at Cold Harbor. At Fredericksburg his unit charged the stone wall three times. After that battle he "...helped to bury the dead in long rows in trenches, laying them one on top of the other, filling up the ditch with pieces of dead men." He somehow survived until war's end. Though ill and exhausted, Deane managed to walk most of the way from Nashville to Chicago (American Historical Society 1922). Clara Barton had aided men from Deane's and other regiments at the battles of 2nd Bull Run, Chantilly, South Mountain, Antietam, Wilderness, Spotsylvania, and Fredericksburg.

During the war Miss Barton had been chosen "Daughter of the 21st Regiment" and was invited to speak at many of their reunions. Nathaniel Deane later described her final appearance at his regimental reunion, on August 23, 1910 when she was 88 years old. He described this "grand old woman" as she feebly rose to speak (American Historical Society 1922 p 59).

> In appearance Miss Barton's face was heavily creased, though it showed few of the smaller wrinkles. She dressed with a kind of careless grace and individuality. Her hair was very dark, with but few gray hairs. She wore a little black poke bonnet and a rich triangular shawl thrown around her shoulders. She wore no jewelry save a pansy cut from a single amethyst presented to her by the Grand Duchess of Baden in memory of their life-long friendship. No matter how simple her attire, this beautiful gem which hung at her throat gave her an air of distinction. Her voice was in the beginning like the aged and this she seemed to realize, for soon she threw herself into the struggle, her eyes seemed to open wide and flash the fire of her intellect, her voice came back to her, and she was again herself, this wonderful woman so near the end of her time.

Much credit is due to those who aided survivors of the 1889 flood, and this is a positive legacy from that terrible event. There were indeed deaths from disease and injury weeks after the flood, and these victims are not included in the "official" death toll. But many more survivors would have perished from exposure, injuries, typhoid, cholera, dysentery, or other illnesses without the diligent work of the sanitary commission and the army of Red Cross workers, led tirelessly by Barton.

On the eve of the Red Cross leaving town the *Johnstown Daily Tribune* (Nov. 1, 1889) published an article titled "Farewell to Miss Barton," which read in part:

> How shall we thank Miss Clara Barton and the Red Cross for the help they have given us? It cannot be done; and if it could, Miss Barton does not want our thanks. She has simply done her duty as she saw it and received her pay —the consciousness of a duty performed to the best of her ability. To see us upon our feet, struggling forward, helping ourselves, caring for the sick and infirm and impoverished—that is enough for Miss Barton. Her idea has been fully worked out, all her plans accomplished. What more could such a woman wish? .... We cannot thank Miss Barton in words. Hunt the dictionaries of all the languages through and you will not find the signs to express our appreciation of her and her work. Try to describe the sunshine. Try to describe the starlight. Words fail, and in dumbness and silence we bow to the idea which brought her here. .... Men are brothers! Yes, and sisters, too, if Miss Barton pleases. The first to come, the last to go, she has indeed been an elder sister to us—nursing, soothing, tending, caring for the stricken ones through a season of distress....

Clara Barton died on April 12, 1912 at her home in Glen Echo, Maryland. This remarkable woman, cherished as the "Angel of the Battlefield," will long be revered as a saint in Johnstown.

14.3   A Walk among the Tombstones

Cemeteries are hallowed ground, especially honored fields of our fallen heroes like those at Arlington, Gettysburg, and Normandy. Visiting a hometown cemetery brings floods of memories of loved ones lost, relatives, and classmates. They are remembered now only in the mind's eye and in photo albums from the last century, many images only in black and white. Through time even the memory of their faces and voices starts to fade, and for some we have lost, all that remains are a few erstwhile pictures and a stone carved with their name and dates of birth and death. I feel sadness for them and the opportunities they never had, for education and travel, as many never left the country and some rarely left the state. They never experienced our amazing world of diverse people, food, music, art, cultures, and vast cities. Many of our ancestors had none of these opportunities as they traveled between the eternities, working hard to live and love, obliged to follow a difficult and often unenlightened path. And some who did travel were sent to war, which would mark them physically and emotionally for the rest of their lives.

There is a cemetery in the hills south of Johnstown, west of the Stonycreek, where over half the victims of that terrible day in 1889 were laid to rest, their life's journeys at an end. At least 430 were buried in private lots, and more than 770 victims who could not be identified or were not claimed by families were interred in a large plot. The flood destroyed many families, orphaned many children, and left others homeless and exposed to the elements in coming days until relief arrived. But those who survived would have another chance at happiness, although absent many family, friends, and home. Grandview Cemetery is a quiet place of peace for these many lost, high on a hill beyond any future flood. Other victims were carried to other graveyards in the region, reflecting their religious faiths. The remains of more than 100 were sent to other towns and States where they had family, including those passing through on the wrong day at the wrong time on trains. Some victims were sent as far as Michigan, Chicago, New York, and New Jersey. The body of Mrs. A. C. Christman was brought to Dallas, Texas, and Annie Bates to Delavan, Wisconsin.

On this warm and beautiful November morning I am meandering through a forest of monuments and mausoleums on a hilltop, not in Johnstown but in Philadelphia. In Johnstown's records and David McCullough's book (1968 p. 277) there was no record of a burial place for Henry Clay Adams, John Parke's fellow student and class President. Henry's father had been a U.S. Congressman from Kentucky who died 5 years before his son, so surely there was a family plot waiting for Henry Clay's remains. Was the family able to recover his body from the chaos of Johnstown? Or was his body placed in the large plot at Grandview or elsewhere?

I learned that Henry's father had moved to Pennsylvania after his time in Congress and after serving as the 6th Auditor of the U.S. Treasury Department, during the war period of 1861 to 1864. He then practiced law in Philadelphia where he remained until his death in 1884. A brief search of Congressional records (2018) revealed he was buried in the enormous West Laurel Hill Cemetery, just west of the Schuylkill Expressway. This historic park has a directory of thousands of burials from prominent families. Sure enough, Greenfield Adams and his wife, Josephine Lippincott (Stokes) Adams, were listed, but not Henry Clay.

Finding the general grave location on a map, I slowly drove through the winding roads at West Laurel Hill. Pulling the car onto the verge I stepped out. Apart from two far-off women walking their dogs the whole place was deserted. Mature trees still kept most of their leaves this late in the season, and they rustled in the slight breeze. The early morning sun cast long shadows from the obelisks. Nearby, a broad bench-like monument bore the name "Adams," but Henry Clay and his parents were not buried here. I looked off to the northeast and saw a group of similar-looking stones, headed by a Celtic ringed cross less than two meters tall, in the design of the ancient high crosses of Britain. This cross bore the names Adams and Stokes. I ducked beneath low-hanging tree branches and came up to the cross. To the right of it were granite markers for John and Eliza Stokes. Then moving left I found them, stones for Henry's mother and father. But was Henry's grave here?

A squirrel in the tree above loosened a stick that crunched into the dry leaves behind me. I twisted round and there, on a marker easily read upside down, was the name Henry Clay Adams. His mother, who eventually lived to see the end of World War I, managed to have her son's body brought here. It would have been unusual in those days to place a stone without a burial, apart from villages of the east where fishermen were lost at sea. Henry's mother had his remains and tombstone placed at the very foot of his father's grave (Fig. 14.1).

All of the original stones in this plot were replaced sometime in the last century. They are identical curved granite markers with raised letters. One minor mystery - 1816 is shown as the birth year of Greenfield Adams. But congressional records say he was born in Kentucky on August 20, 1812. Perhaps the original stone was marble

Fig. 14.1 Gravestone of Henry Clay Adams, who died in the Johnstown flood

and growing illegible, which led Adams' descendants to thoughtfully replace the markers. Acid rain, largely from the burning of coal and other fossil fuels, severely degrades the lettering on old marble sculptures and tombstones.

After photographing other Adams' markers I strolled around to check out other names from that bygone era. Across the road was a long, low monument shaped like a casket, with a wide periphery of grass with no markers. Curious about this strange plot, I walked over and was surprised to read the names of John Reilly and relatives. Surely this was not the same John Reilly, Congressman from Altoona (1875–77), who sold the South Fork dam to the Fishing and Hunting Club in 1880. The modern era of the smart phone now proved its worth in a quick internet search. After he sold the South Fork dam property, Reilly moved to Philadelphia in 1881 as part of his railroad work and lived there until he passed in 1904. The Congressional record revealed his place of burial - here at West Laurel Hill! He was born February 22, 1836. The dates for his birth and death appear as Roman numerals on the monument: MDCCCXXXV (1835) and MCMIV (1904). His wife's name was Anna Lloyd Reilly (Jordan 1921), the same as on the marker. Despite a birth year error, the stone carver needing a 10th digit, this is indeed the resting place of Congressman Reilly of Altoona.

The remains of young Henry Clay Adams, drowned in the Johnstown flood, and Congressman John Reilly, former owner of the South Fork dam, lie only 30 meters apart on a hilltop in Philadelphia.

14.4　Uldis Kaktins, in Memorium

Dr. Uldis Kaktins, Emeritus Professor of geology, was born on June 10th, 1942 to Zigurds and Zenta Kaktins in Riga, Latvia during the heart of World War II. When he was about 2 years old, Uldis' family fled Latvia to escape Stalin's invading

Russian troops and deportation to Siberia. Five years of his childhood were spent in displaced persons camps in Austria and post-war Germany while his mother and father struggled to move their family to the United States, eventually arriving in Boston. Uldis remembered drinking a glass of fresh milk and eating a donut, a very fond memory for him! He grew up in a Boston ghetto, and Uldis joked about being picked on because he was short, had a funny name, and didn't speak English. But the family put down new roots and Uldis worked hard to get an education.

The Vietnam War interrupted graduate school for geology. Uldis was deployed to Vietnam and wrote his thesis by hand on the floor of his Bachelor Officer's Quarters during the Tet II Offensive. He eventually earned his PhD from Boston University and worked as a professor and chairperson of the geology department at the University of Pittsburgh at Johnstown, where he taught from 1975–2008. Professor Kaktins researched the floods of Johnstown throughout his career and into retirement, culminating in published papers in 2013 and 2016 and a chapter in the field guide for the 2016 Field Conference of Pennsylvania Geologists. Uldis taught and mentored countless students, many of whom became close family friends. He passed away on Saturday, July 2, 2016, a three-year survivor of pancreatic cancer. His academic career and military service are wonderful examples of the richness that immigrants bring with them to start new lives in the United States. His legacy lives on through his family, his research, this book, and with the geologists he inspired who ply their trade around the globe from Johnstown to Singapore (Fig. 14.2).

Fig. 14.2 Professor Emeritus Kaktins on the "Path of the Flood" hiking trail west of South Fork, near the viaduct bridge (*left*), and reconnoitering a mountain stream (*right*)

14.5   Dam Safety – A Terrible Lesson from the Past

In early August of 1975 in Hunan province, Chinese fishermen, farmers, and villagers along the Hong and Ru Rivers awoke to another day of struggle against the elements. For days the remnants of Typhoon Nina had soaked the region with torrential rains. As the storms began to diminish the people believed the worst was over. But it wasn't so. In the wee hours of August 8th two large dams on the Hong and the Ru were filled to their crests by runoff that they could not safely discharge. Both dams overtopped and breached, releasing massive flood waves into the river valleys. The "River Dragon" had come! (Dai Qing 1998).

Within hours more than 70,000 people would be swept to their deaths. Eleven million would be displaced from their homes. Altogether, 230,000 are thought to have perished, mostly from disease and famine after the flooding. Yi Si (1998) describes the tragedy and gives a compelling history of the fatal dams.

I personally had a taste of the cyclones in that region. Nine months earlier I had been hunkered down in a Quonset hut in the Philippines as a typhoon lashed the island of Luzon. You can never forget the crashing of trees or the deafening thunder of coconuts, wrenched by the wind from palm trees and hurled against a metal building.

Super Typhoon Nina had begun as a disturbance in the Philippine Sea, strengthening rapidly to tropical storm and then to hurricane strength. Reconnaissance aircraft experienced sustained winds of 155 mph by August 2, 1975. The storm moved northwest, weakened somewhat, and made landfall in Taiwan as a Category 3 storm. The weakened system crossed the Formosa Straits, struck Jinjiang on the Chinese coast, then tracked farther northwest into mainland China.

Then catastrophe struck – a cold front blocked the storm, which strengthened and became a stationary system of thunderstorms with three periods of heavy rain. Upwards of one meter of rain fell in 3 days on the basin of the Huai River (Yi Si 1998). Massive runoff surged into two large reservoirs on tributaries of this river, the Shimantan Dam on the Hong River, and to the south the larger Banqiao Dam on the Ru River, Henan Province (built 1951–52 with Soviet consultants). Both dams failed catastrophically. The peak discharge through the breach in Banqiao Dam was estimated at $78,100 \text{ m}^3 \text{ s}^{-1}$. Downstream were scores of smaller dams and levees in the paths of the flood waves. Altogether, the two large dams, two medium dams, Tiangang and Zhugou, and 58 small dams failed from overtopping (Xu et al. 2008). Out of desperation officials ordered some of the embankments downstream to be bombed so that the water in those impoundments could be released before the larger flood waves arrived. Large lakes had formed trapping hundreds of thousands of people on small "islands" of higher ground. Other embankments were destroyed with explosives to release those lakes (Yi Si 1998).

The Huai had always been terribly flood prone. After destructive flooding in 1950, the government announced a campaign to "Harness the Huai River" (Yi Si 1998). Their intention was good - to tame the river by placing large dams on its main tributaries and constructing levees and catchment areas downstream to support agriculture. The Banqiao and Shimantan reservoirs would also generate needed hydroelectric power for the region. Banqiao Dam was thought so strong as to be invulnerable – some called it an "iron dam" (Yi Si 1998). But the design of both dams was terribly unsound – they did not have enough discharge gates to safely pass the runoff from exceptionally large storms. Typhoon Nina brutally exposed this flaw. Both dams were overtopped, the worst fate that can befall earthen dams. The South Fork Dam suffered the same fate in 1889.

For many years the true nature of the 1975 disaster in China and the deaths that resulted had been downplayed. The Chinese people have experienced horrendous floods for centuries, indeed for millennia. The Hwang Ho, or Yellow River, has been the most destructive river in the world. Flooding on the Hwang Ho in 1887, 2 years before the Johnstown flood, killed nearly two million people. Upwards of four million died in 1931 (O'Connor and Costa 2004), then flooding again in 1938 killed almost one million people (Hudec 1996). These enormous death tolls were a result of dense populations and rapid growth in flood prone areas susceptible to extreme tropical storms. Regional floods appear to be enhanced in El Niño years.

But the 1975 destruction in China was largely caused by the collapse of engineered works, the sequential "domino" failure of dozens of dams and levees that had been intended to harness and control the river systems, not lay waste to an entire region. The death toll was exacerbated by the lack of an effective warning system. Ill-conceived cost cutting measures in the end cost many lives. The government had been cautioned that more discharge gates were needed on the large dams and had also been advised not to develop some of the long-standing floodplains. The financial cost of the destruction and rebuilding of dams and levees was many times greater than the added cost that would have been incurred to include more discharge gates. And no price can be placed on the precious lives lost.

References

American Historical Society, Inc (1922) History of Pittsburgh and environs, pp 57–60

Barton, CH (1910) The Red Cross – in peace and war, American Historical Press, p 703

Hudec K (1996) NOVA: dealing with the deluge. Posted March 26, 1996. Available at: http://www.pbs.org/wgbh/nova/earth/dealing-deluge.html (Accessed 27 Apr 2018)

Johnstown Daily Tribune (1889) "Farewell to Miss Barton". 1 Nov 1889

Johnstown Daily Tribune (1890) "Looking for a Lost Brother". Wednesday, June 25, p 1

Jordan JW (1921) Encyclopedia of Pennsylvania biography (illustrated with portraits). Lewis Historical Publishing Company, New York, pp 82–84

McCullough DG (1968) The Johnstown flood. Simon & Schuster, New York, 302

O'Connor JE, Costa JE (2004) The world's largest floods, past and present—their causes and magnitudes: US Geol. Survey Circular 1254, p 13

Qing D (1998) The river Dragon has come! The three gorges dam and the fate of China's Yangtze river and its people. Editors: Thibodeau JG and Williams PB

Si Y (1998) The World's most catastrophic dam failures: The August 1975 collapse of the Banqiao and Shimantan dams, Chap. 3 in Dai Qing, *The River Dragon Has Come!*, M.E. Sharpe, NY

The Daily Tribune - Johnstown (1900) "Flood Victim Turns Up". Saturday, Friday, 18 August
    1900, p 5
US Congress (2018) Biographical directory, Adams, .Green, 1812–1844 http://bioguide.congress.
    gov/scripts/biodisplay.pl?index=A000036 (Accessed March 2018)
Xu Y, Zhang L, Jia J (2008) Lessons from Catastrophic dam failures in August 1975 in Zhumadian,
    China. In proc., GEOCONGRESS 2008: Geosustainability and Geohazard Mitigation,
    pp 162–169

Epilogue

Johnstown changed after the 1889 flood. It became a city through the merger of the former boroughs, some of which had been virtually wiped out. The story has yet to be well told of how the survivors came together to quickly rebuild their lives, homes, and businesses, although many chose higher ground. The liability laws in Pennsylvania and other states soon changed to more reflect those in England. Ever after, those responsible for calamities would be hauled into the courts, though even then with predictably uneven results depending on the wealth and stature of the defendants - people and corporations. Justice is supposed to be blind, and at its best our system serves as a model for the world, designed for self-repair by thoughtful patriots, the authors of the Constitution. Let us not forget that in some States of the U.S., as recently as the 1960's, it was a felony for a man and a woman to marry, if they happened to be of different races.

Some responsibility will always lie with those who build in places where floods, earthquakes, landslides, and hurricanes have struck in the past. But our people lead very busy lives, and present generations may not recall decades old catastrophic events. It then becomes the certain responsibility of officials who approve new construction, those who have the maps of past areas of inundation and destruction, to guide the public and real estate developers about where it is safe to build, and to deter development where there are great risks. There are few human endeavors with greater consequences for ignoring the past.

The engineering investigation of the Johnstown flood of 1889 is a tale of powerful men seeking to protect their reputations, then and now, by hijacking a safety investigation and delaying and distorting its results. The moral dilemma lives with us still, the senseless but omnipresent conflict of power versus truth. But in matters of public safety, when the lives of thousands and perhaps millions may hang in the balance, truth must somehow rise above the chaos that power and greed all too often inflict upon it.

© Springer International Publishing AG, part of Springer Nature 2019
N. M. Coleman, *Johnstown's Flood of 1889*,
https://doi.org/10.1007/978-3-319-95216-1

Appendices

Appendix 1: Table of Precipitation Data

Precipitation (inches) May 30 and May 31, 1889 measured at Pennsylvania stations. Note that Johnstown is not included because the gage was swept away shortly after the last reading at 10:44 am; the total up to that time had been 2.0–2.3 inches. Blodget (1890) estimated that the storm total for Johnstown was 3.0–3.5 inches. (Sources: Blodget 1890; Russell 1889; Townsend 1890) [1.00 in = 2.54 cm].

Station	Total	Comments
Altoona	5.33	Began 3:30 pm 30th, ended 9 pm 31st (36 km NE of south Fork dam)
Bethlehem	0.24	
Blue Knob	7.90	Began 3:20 pm 30th, ended 9 pm 31st (19 km E of South Fork dam)
Brookville	2.41	Began 1:30 pm 30th
Carlisle	2.45	
Charlesville	7.60	Began 3:15 pm 30th
Clarion	3.02	Began 2 pm 30th, ended 11 am 31st
Columbus	0.83	
Confluence	1.14	Began 4:30 pm 30th, ended 7 pm 31st (76 km SW of South Fork dam)
Corry	0.81	
Coudersport	5.40	Began 5 pm 30th
Dyberry	0.46	Minimum
Eagle's Mere	5.53	
Emporium	5.97	Began 5 pm 30th, ended 11 pm 31st
Erie	0.92	
Franklin	0.91	
Freeport	1.80	Began 3 pm 30th
Germantown	0.51	
Girardville	1.54	
Grampian	8.60	Began 4:30 pm 30th, ended 11:20 pm 31st
Greensburg	1.70	Began 6 pm 30th

© Springer International Publishing AG, part of Springer Nature 2019 229
N. M. Coleman, *Johnstown's Flood of 1889*,
https://doi.org/10.1007/978-3-319-95216-1

Station	Total	Comments
Greenville	1.22	
Harrisburg	4.86	Minimum, doesn't include some rain night of 31st
Hollidaysburg	5.71	Began 4 pm 30th, 5.51″ by 9 pm 31st, ended 12 pm 31st (34 km E of South Fork dam)
Honesdale	0.57	
Huntingdon	6.01	Began 4 pm 30th, ended 2 am June 1st
Indiana	3.00	Ended 7:30 pm 31st (44 km NW of south Fork dam)
Lancaster	0.73	Minimum
Lebanon	2.52	
Leroy	2.08	
Mahoning	0.64	
Marshalls Creek	0.60	Monroe Co., near Stroudsburg
Meadville	1.35	
Mercersburg	1.41	Minimum?
Myerstown	1.75	
McConnellsburg	8.99	Began 4 pm 30th
Neshaminy	0.54	
New Bloomfield	4.19	
New Castle	1.30	
Nisbet	3.10	Began 3 pm 30th
Oil City	1.03	
Ottsville	0.53	
Parker's Landing	1.02	
Petersburg	6.61	Began 3 pm 30th
Phillipsburg	3.33	Began 3:50 pm 30th (76 km NE of south Fork dam)
Pittsburgh	1.49	
Point Pleasant	0.51	
Pottstown	1.70	
Saltsburg	1.94	
Selinsgrove	6.46	
Smethport	5.50	Began 11 pm 30th
Somerset	4.43	Began 10 pm 30th (45 km SSW of south Fork dam)
State College	3.10	Began 3:45 pm 30th
Tipton	4.15	Began 4 pm 30th
Tuscarora	5.81	Juniata co.
Uniontown	2.07	Began 8:50 pm 30th, ended 8:30 am 31st
Warren	1.76	Began 6 pm 30th
Wellsboro	2.35	
West Chester	0.50	
Wysox	3.22	
York	1.51	

Appendix 2: Establishing a GPS "benchmark" at the South Fork Dam, and Finding the Difference between the 1889 and Modern GPS Survey Elevations

GPS Benchmark

In a contribution to this research, Musser Engineering of Central City, PA established an RTK (Real Time Kinetic) GPS base station near the South Fork dam. This GPS "benchmark" greatly facilitates our research at the dam site and is documented here to support future investigators. This base station is east of the dam beside the footpath that leads from the eastern parking area to the bridge over the spillway. The station is on a circular concrete base with a preexisting brass plate (Fig. A2.1). Data for this station are given in Table A2.1. Musser Engineering used an RTK GPS unit at the base station, a TOPCON HiPer Ga, manufactured by Hayes Instrument Co. A laser theodolite was used at the established base to determine coordinates for other features at the site. The Total Station Instrument was a TOPCON GPT 3005 W. The absolute elevations obtained by Musser Engineering were further used to calibrate relative GPS measurements collected at the site by Professor Brian Houston of the UPJ (University of Pittsburgh at Johnstown) Engineering Department (Fig. A2.2). To convert these relative data to NAVD 88 (North American Vertical Datum of 1988) we added 13.1 ft. The common point in these two GPS surveys was a foundation stone at the base of the dam (Fig. A2.2). Note that Figs. A2.1 and A2.2 were originally published as supplementary online material for the paper Coleman et al. (2016).

Difference between the 1889 Survey Elevations and Modern GPS Data

LiDAR data and results acquired in two GPS surveys at the dam site were used to evaluate the difference between the 1889 and modern-day elevation reference frames. It was challenging to find sites where the surfaces have been relatively

Fig. A2.1 Photos showing the RTK GPS "benchmark" location with concrete base and brass plate. This was a pre-existing Park Service marker, with the plate labeled "NPS 84 JOFL." Interstate 219 bridge is visible through the dam breach at upper left. Photos by the author

Table A2.1 Altitude Data for the South Fork Dam (Sources: Musser Engineering, Central City, PA; UPJ Engineering Dept.; and PASDA, 2013a for LiDAR data)

Location	Altitude NAVD 88 (msl)	Latitude (or Northing)	Longitude (or Easting)	Notes/source
RTK base at JOFL #84	485.606 m (±0.025 m) 1593.1920 ft	N40° 20′ 52.74434″	W78° 46′ 21.93709″	Converted to NAVD 88 using GEOID09 by Musser Eng.
Stone at base of former control tower	1544.93 ft			Musser Eng. Laser theodolite
Natural surface on eastern dam abutment	1620.90 ft			Musser Eng. Laser theodolite
Spillway top of soil surface near 3rd bridge support post from E end	1608.8 ft. (1595.66 ft. + 13.1 ft)	Northing 371685.331	Easting 1683262.211	UPJ engineering GPS
Natural surface on west side of spillway footbridge	1620.8 ft. (1607.738 ft. +13.1′)	Northing 371766.455	Easting 1683177.966	UPJ engineering GPS
Soil surface in area of main spillway crest	1611 to 1612 ft			Pennsylvania LiDAR 3.2 ft. DEM (see Fig. A2.3 for map area and scale)

stable since 1889. Three places at the dam were chosen where there is confidence the surfaces are little changed, including: (1) the bedrock spillway surface beneath the soil layer, (2) station 1050 from the 1889 topographic survey, and (3) the highest points on the western dam remnant.

The spillway surface would seem to be the best location, since it was a bare rock surface in 1889. However, a post-flood photo of the spillway shows that it had an uneven surface with scattered rocks (Francis et al. 1891 Plate XLVI), and over >120 years a soil of leaf litter and colluvium has accumulated. The ASCE topographic survey took measurements directly on the spillway surface. "By our levels [measurements], the floor of the wasteway for 176 feet from the lake averages 1602.82 feet above tidewater" (Francis et al. 1891, p 446). The locations and number of measurements averaged by the committee were not shown in their report, but they noted the range of elevations varied from 0.14 ft. higher to 0.08 ft. less than 1602.82 ft. They did provide a diagram (top of their Plate LI) showing the profile along the axis of the spillway, with a base level of 1602 ft. It is unclear why they show this number instead of rounding up their measurements to 1603 ft. Another estimate for the spillway floor was given by P. F. Brendlinger, who took independent measurements a year after the dam breach and gave a spillway elevation of 1603.4 ft.

Fig. A2.2 Location of GPS measurement (UPJ engineering) on top of foundation stone at center of former dam. These dressed stones formed part of foundation for the original control tower from which 5 large discharge pipes were operated. Note partly submerged remains of logs at bottom right. Hemlock logs were placed by the South Fork Fishing & Hunting Club as part of their dam repair and modification. Photo by the author

(Francis et al. 1891). This was not the spillway surface but was the "top of the [bridge] sill" at a location 26 ft. (7.9 m) from the dam end of the bridge.

LiDAR data (PASDA 2013b) for the surface of the main spillway show that it transitions from ~1609 ft. at the upstream entry to a crest of ~1612 ft. (491.3 m, NAVD 88) over a distance less than 20 meters (Fig. A2.3). To augment the LiDAR digital elevation model (DEM), which has a vertical accuracy of <1 m, a high-resolution GPS measurement is available for the soil *surface*, at a point 1 m south of the 3rd bridge support post from the eastern end. That GPS elevation is 1608.8 ft. (490.36 m, NAVD 88). As part of a 2011 soil survey permit (NPS 2012) we used a thin penetrator rod to measure soil thicknesses at several points along the axis of the spillway. No soil samples were taken, only depths were determined. The depth of refusal for the thin rod represents a minimum soil thickness. These measurements are used to estimate the elevation of the subsoil spillway surface. At the GPS location the soil thickness is ≥27 cm. Subtracting this, the elevation of the spillway rock surface is ≤490.1 m (NAVD 88). However, this location is near the spillway "lip" and not near the crest of the spillway. Starting from the GPS location, at a distance 35 m to the north along azimuth 12°, three penetrator measurements 0.1 m apart reveal soil depths of 53 to 56 cm. The narrow range suggests the soil was fully penetrated. The surface elevation here is ~490.9 m, determined with an inclinometer in two stages over the 35 m distance. Subtracting the soil depths, the spillway surface at the crest has an elevation of ~490.4 m (NAVD 88). The result is a difference of

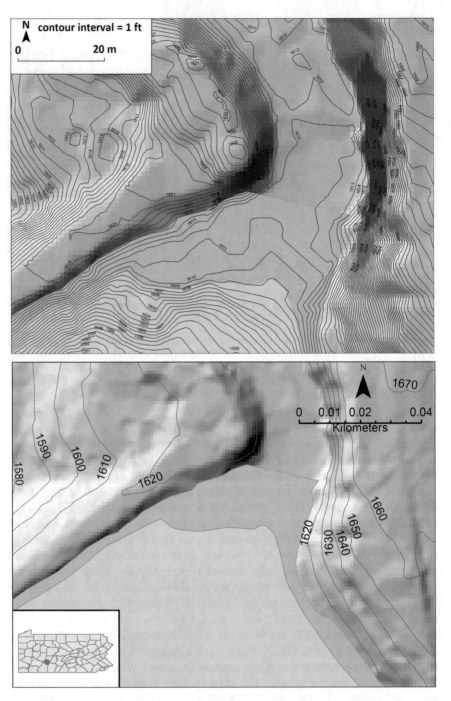

Fig. A2.3 Contour maps at two scales showing main spillway at northeast end of dam. Flow direction was from bottom to top of frames. Produced using LiDAR 3.2 ft. DEM of Pennsylvania (PASDA 2013b). Bedrock excavated from this and the auxiliary spillway was one source of the riprap used to cover dam slopes. Upper panel shows contours for the entry and central portion of spillway floor. Elevation data for the DEM were obtained before Dec. 2006. The present-day bridge across spillway is not shown (credit: S. Wojno)

6.1 ft. (1.9 m) between the 1889 survey and modern GPS-based data. Compared to the more accurate GPS measurement, the LiDAR DEM over-predicts the soil surface elevation at the spillway crest by about 0.3 m, which is within the DEM accuracy range. Part of this small over-prediction might be influenced by conditions in the spillway itself, including its vegetated surface, tree-lined upper margins, and steep side slopes. Reports of vertical accuracy testing for LiDAR data published for Pennsylvania are provided by DCNR (2018).

The second location for elevation comparison is on the eastern dam remnant, near the site of an 1889 measurement point, station 1050 (see Francis et al. 1891 Plate L), for which an elevation of 1613.34 ft. was reported. A modern GPS measurement near this point yields 1619.44 ft. (NAVD 88), equivalent to 493.60 m. The difference between the 1889 and modern reference frames is ~1.86 m (6.10 ft). This location appears little changed since 1889, because this is the intersection of the embankment with natural ground on the eastern abutment. About 6 m to the northeast a vestige of the old road from South Fork intersects the eastern abutment. High points on the western dam remnant comprise the third site for elevation comparison. The elevation data show that the surface of the footpath on the crest of the western remnant has been leveled since 1889. The three highest points from the 1889 survey are ASCE stations 250, 300, and 325, with elevations in the range 1611.14 to 1611.24, varying only 0.1 ft. (3 cm). Two GPS measurements on this same section of crest yield 1617.49 ft. (NAVD 88). The difference between the 1889 and GPS reference frames is ~6.3 ft.

Based on the close agreement of measurements from all three locations, we are confident that the difference between the 1889 and present-day reference frames is in the range from 6.1 to 6.3 ft., with 6.2 ft. (~1.9 m) being our best estimate. This difference was used to construct the dam remnant profiles in Chap. 9, Fig. 9.2.

Appendix 3: Estimating the Time of Concentration (t_c) and Time to Peak (t_p) for the South Fork of the Little Conemaugh River

To estimate the time of concentration (t_c) for the South Fork basin and time to peak discharge (t_p) on the day of the flood, multiple methods were applied that have been developed or used in various parts of the U.S. When applicable, the Kerby-Kirpich method (Roussel et al. 2005) can be used to estimate t_c. The velocity method (NRCS 2010) is also commonly used. Both methods estimate t_c as the sum of travel times for discrete flow regimes. The Kerby-Kirpich approach requires few input parameters, is straightforward to apply, and produces readily interpretable results. Resulting t_c estimates are consistent with watershed time values independently developed from real-world storms and their runoff hydrographs (Roussel et al. 2005). Fang et al. (2008) also report that the Kirpich and Haktanir–Sezen methods provide reliable estimates of mean values of t_c variations. Overland flow in the uplands, especially the higher altitude areas of the eastern South Fork watershed, would significantly add to the magnitude of t_c. Therefore, rather than using the Kirpich equation, the Kerby-Kirpich method is preferred because it yields a total t_c by adding the overland flow time (Kerby) to the channel flow time (Kirpich):

$$t_c = t_{ov} + t_{ch}$$

where:
t_c = time of concentration.
t_{ov} = overland flow time.
t_{ch} = channel flow time.

Roussel et al. (2005) analyzed 92 watersheds in Texas, and examined various methods for evaluating the time-response of watersheds to rainfall. They emphasized that none of the watersheds in their database had low topographic slopes. They recommended the Kerby-Kirpich approach as a preferred way to estimate t_c. Their study included watershed areas ranging from 0.65 km^2 to 390 km^2, main channel lengths between 1.6 km and 80 km, and main channel slopes between 0.002 and 0.02 (Roussel et al. 2005). The main channel slope is the change in elevation from the watershed divide to the outlet divided by the length of the main channel and tributaries leading from the divide. For smaller watersheds and intense storms where overland flow is an important component, the Kerby equation can be used:

$$t_{ov} = K\left(L \times N\right)^{0.467} S^{-0.235}$$

where:
t_{ov} = overland flow time of concentration (minutes)
K = a units conversion factor; 0.828 for traditional units and 1.44 for SI units
L = the overland-flow length (feet or meters)
N = a dimensionless retardance coefficient
S = the dimensionless slope of terrain conveying the overland flow

For the most distant parts of the South Fork watershed at upland flow divides, the length of overland flow is estimated at ~600 m. A dimensionless retardance coefficient (N) of 0.8 was chosen to represent deciduous forest with thick leaf litter. The slope of the overland flow component for uplands near the drainage divides varies considerably and was estimated to range from 0.1 (i.e., 60 m/600 m) to 0.01. Given these values, the Kerby equation yields a t_{ov} range from 44 to 76 min. These values exceed the 30-min period that is sometimes treated as a "standard" lag time for overland flow.

The Kirpich equation is used to compute the channel-flow component of runoff:

$$t_{ch} = K\left(L\right)^{0.77} S^{-0.38}$$

where:

t_{ch} = the channel-flow component of time of concentration (min)

K = a units conversion factor; 0.0078 for English units and 0.0195 for SI units

L = the channel flow length (feet or meters), or total basin length minus overland flow length

S = the dimensionless main-channel slope

The longest channel system in the drainage basin is the main channel of the South Fork of the Little Conemaugh and its tributaries that reach to the southeastern corner of the basin. When calculated using a series of straight-line segments, the main channel is ~17 km long. The extra length of small-scale meanders was not included, because at high flood levels water overflows the banks and follows a more direct path. A mean channel slope of 0.016 is obtained by dividing the elevation difference between the lake basin and the higher channel reaches to the southeast (i.e., 270 m) by the channel path length (17 km). These values yield a t_{ch} of 170 min. Adding the t_{ov} range of 44 to 76 min gives a t_c range of 214 to 246 min, or 3.6 to 4.1 h.

For comparison, the Haktanir and Sezen (1990) method was applied (see Roussel et al. 2005) and yielded a larger t_c = 4.9 h. The Folmar and Miller (2008) equation (also see NRCS 2010) is based on the longest hydraulic length in the basin, and was developed using 10,000 direct runoff events from 52 agricultural watersheds. The watersheds ranged in size from 3 acres up to 20 mi^2. They found that lag time correlated strongly with the longest hydraulic length. When applied to the larger South Fork watershed, their equation yields a lag time of 6.7 h, with lag time being measured from the centroid of excess precipitation to the discharge peak on the hydrograph.

An additional method was applied to estimate t_c (in hours) for a basin, by taking the square root of the basin area in *square miles* (Roussel et al. 2005; Fang et al. 2008). This method has no apparent physical basis; but, when graphed with a large number of t_c values derived from a group of observed rainfall-runoff analyses, the resulting reference line passes through the general center of the data plot (Fang et al. 2008). The South Fork basin has a total area of 160 km^2, but the drainage area above the former lake basin is less, about 137 km^2 (53 mi^2). The square root of 53 yields a t_c of 7.3 h. I consider that this value, with the Kerby-Kirpich and other results, provides a practical t_c range of 3.6 to 7.3 h.

Roussel et al. (2005) show that t_c can be used to estimate the time to peak discharge (t_p). When the Kerby-Kirpich approach is used, for undeveloped watersheds they advise using the relation $t_p \approx 0.7 \, t_c$. Applying this empirical approach to the South Fork watershed yields an estimated t_p range from 2.5 to 5.1 h. This relatively small magnitude of t_p is credible for the compact watershed of the South Fork of the Little Conemaugh.

Stream gaging data are not yet available for the South Fork of the Little Conemaugh River,[1] but due to its frequency and severity of flooding, long-period

---

[1] As of 2018, gaging data for this stream are being collected by researchers from the Dept. of Energy & Earth Resources, University of Pittsburgh at Johnstown, in cooperation with the National Park Service.

data have been collected for the Little Conemaugh itself (NWIS 2015). The river has a drainage area of 482 km² upstream of the USGS gage at East Conemaugh, Pennsylvania. In response to most regional storms, the time of concentration, time to peak, and lag time must be significantly less for the South Fork subbasin, which is far upstream from the gage and has a much smaller drainage area. Very large runoff events have been documented for the Little Conemaugh River, the largest being the July 20, 1977 flood with a peak discharge of ~40,000 cfs determined at East Conemaugh (Brua 1978). In a 1040-km² area north and east of Johnstown, rainfall totals of 6 to 12 inches were measured over 6 to 8 h. Mapping of rainfall showed that the greatest amounts fell over the South Fork and the Laurel Run sub-basins. Forty people were killed when Laurel Run dam #2 failed in 1977 (Brua 1978 Fig. 3). Six other dams in the region also failed. Brua (1978) estimated a peak discharge in the South Fork of the Little Conemaugh at 24,000 cfs, or 680 m³ s⁻¹ (at Fishertown, just below the former South Fork dam). He also estimated a unit discharge of 456 cfs/mi² (5.0 m³ s⁻¹ per km²) for this subbasin. Brua's (1978) compilation and analysis shows great variability of unit discharge over the subdrainages of the Little Conemaugh River and demonstrates the severe challenge of identifying "design storms" for engineering purposes in basins of this size. The South Fork sub-basin makes up only 28% of the Little Conemaugh basin by area, but in the 1977 storm it produced an estimated peak discharge (Q_p) >60% of the Little Conemaugh River Q_p as measured at East Conemaugh.

Some insights about time to peak for the South Fork channel can be gleaned from the Little Conemaugh gaging data. Land use, terrain slopes near the divides, and the approximate proportions of forested and cultivated lands are similar for the South Fork subbasin and the basin of the main stem, or North Fork of the Little Conemaugh. The streams respond quickly to precipitation events, so that the time from the initial stream response to the discharge peak is relatively short. This time span is a useful approximation of the lag time, which is the interval from the excess precipitation centroid to the peak river discharge. In the extraordinary flood event of 1977, the time to Q_p following initial river stage response was <10 hr. for the Little Conemaugh at East Conemaugh (Fig. A3.1). The East Conemaugh gage is ~20 km downstream from the former South Fork dam. Therefore the initial response of the stream to precipitation should have been quicker in the smaller subbasin above the dam and t_p must have been significantly shorter (< 8 hr., and probably even less given the estimated range of t_c). This time to peak should represent a reasonable upper limit for the response of the South Fork subbasin to the lesser precipitation amounts in the 1889 storm. Therefore, if the rainfall peak (referring to centroid of precipitation) on the night of May 30–31 occurred between 3 and 6 a.m., then local stream discharges would have been expected to peak and start diminishing between 11 a.m. and 2 p.m. (or earlier) on May 31st. That expectation is consistent with local stream observations around noon on that date. There is also an historical record of river levels measured at Johnstown on the morning of the 1889 flood. At 7:44 a.m. on May 31st, the river level was already 13 ft. higher than the measurement taken 24 h earlier and was continuing to rise. That information is discussed in Chap. 9.

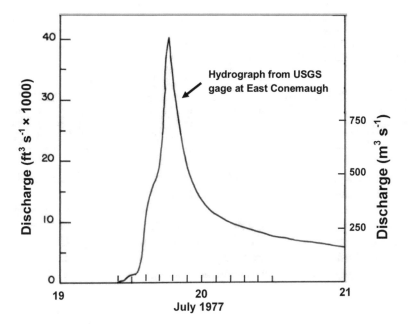

Fig. A3.1 Discharge hydrograph for the Little Conemaugh River at East Conemaugh during the Johnstown flood of 1977 (after Brua 1978 Fig. 20). In this extreme runoff event, the discharge rate at this gage downstream on the main stem of the river exceeded 50% of its peak for only 5 h

Additional hydrologic research about the Little Conemaugh system was published by Roland and Stuckey (2008). They developed regression equations to estimate flood discharges at various recurrence intervals for ungaged streams in Pennsylvania with drainage areas less than 2000 square miles. Their equations were derived with gaging data from 322 monitoring stations in Pennsylvania and adjacent states. Roland and Stuckey (2008) noted that the regression equations are not valid for larger drainage basins, for basins with substantial mining activity, or those with streams regulated by dams. The basin of the Little Conemaugh River has several dams with limited regulation and extensive former underground mining that could affect base flow conditions but would not likely affect storm runoff. This basin has a long period of gaging, dating back to 1936, and these data were used to describe the recurrence intervals for the flood discharges in Table A3.1. It is unclear how the 500-yr *observed* estimate was derived for the table. The highest discharge rate recorded at East Conemaugh in the last 80 years was 40,000 cfs during the Johnstown flood of 1977 (Brua 1978), which included extraordinary rainfall totals of 11 to 12 in (28 to 30 cm) in the South Fork drainage basin.

In summary, the above analysis and discussion yields a range of t_c for the South Fork subbasin of 3.6 to 7.3 h, and t_p (similar to lag time) in the range of 2.5 to 5.1 h. Data for the 1977 flood from the stream gage at East Conemaugh suggest that, in

Table A3.1 Little Conemaugh River flood-flow magnitudes for various recurrence intervals computed from observed streamflow-gaging data at East Conemaugh, with predictions from regional regression equations and a weighted average for gaging stations used in the analysis (reprinted from Roland and Stuckey 2008 p 50)

USGS gaging station no.	Flood-flow estimates (cfs) [1 m^3 s^{-1} = 35.3 cfs]						
	Type	2 yr	5 yr	10 yr	50 yr	100 yr	500 yr
03041000	Observed	10,400	16,800	22,100	36,800	44,500	66,500
(E Conemaugh)	Predicted	11,700	17,900	22,600	35,300	41,700	59,500
	Weighted	10,400	16,900	22,100	36,600	44,100	65,700

extreme events upstream on the South Fork, t_p would be <8 h. And since t_p will usually be less than t_c, it is likely that t_p would be less than 7.3 h.[2]

If the heaviest rainfall over the South Fork basin occurred in the pre-dawn hours, then inflows to Lake Conemaugh should have peaked around noon on the day of the flood. But in repairing the dam the Club had lowered its top – by 11:30 a.m. floodwater had filled the reservoir to the top of the dam and water was surging over the crest, starting to erode the downstream face. The dam was now doomed.

Appendix 4: "The Lesson of Conemaugh"

Major John Wesley Powell wrote an essay titled "The Lesson of Conemaugh," in response to the 1889 disaster in Johnstown. Powell was a celebrated geologist and explorer who in 1881 became the energetic second director of the U.S. Geological Survey, which had been formed by consolidating several western Surveys. There could have been no better advocate for this fledgling agency whose work would be so important for the geologic exploration of the U.S., particularly of the western regions, and for the development and protection of natural resources.

Books about the life and adventures of Powell include those by Worster (2001) and Dolnick (2002) and by Powell himself (1875). Also see the information book by the USGS (1976). As a Union officer during the Civil War, Powell lost his right arm to a rifle bullet at the Battle of Shiloh. He became a national hero in 1869 when, leading a small party of men that included his brother Walter, he entered the Green River in Wyoming with four wooden boats. From the Green River they floated down to the Colorado, and traversed its rock-studded channels and treacherous rapids through the Grand Canyon in the three surviving boats. They emerged 3 months later on August 30th near the mouth of the Virgin River in what is today Nevada. The canyon was so isolated that no communication was possible. Their party had been given up for dead. Three members of Powell's party (William Dunn and the Howland brothers) had left the expedition just 2 days before it ended. They vanished, and were never heard from again.

---

[2] The time to peak discharge can exceed the theoretical time of concentration during protracted storms, especially when successive bands of increasing precipitation move across a drainage basin.

Powell's essay was published less than 3 months after the 1889 flood and nearly 2 years before the release of ASCE's investigation report. Therefore his essay did not include insights from that investigation. I also found no evidence that he visited the scene of the dam breach before writing the essay. His wide-ranging themes strayed far beyond the Johnstown disaster, touching on topics he was advocating to the broader federal government. Powell foresaw generations of dam building for the purposes of flood protection, supplies of safe drinking water, irrigation of arid lands, and power generation. He wrote:

> ...it is probable that the resort to water-power will rapidly increase in the immediate future. It certainly will if the dream of modern electrical science is realized, so that water-power can he economically converted into electric power....
>
> ...flood waters must be stored and allowed to find their way to the sea during times of low water.
>
> Wherever the houses of men are clustered reservoirs or systems of reservoirs must be built.
>
> About two-fifths of the area of the United States is so arid that agriculture is impossible without artificial irrigation, the rainfall being insufficient for the fertilization of ordinary crops. ... In all of this country, wherever agriculture is prosecuted, dams must be constructed, and the waters spread upon the lands through the agency of canals.

One of Powell's prognostications has not quite come true, in which he predicted that our deserts would become virtual utopian gardens:

> Ultimately one of the great agricultural regions of this country will be found in the irrigated plains and valleys of the West. Sage-brush plains, sand-dune deserts, and alkaline valleys will be covered by gardens, fields, and groves, all perennially fertilized from thousands of mountain lakes.

In this Powell overestimated the amount of water that could be stored in and transported from mountain lakes and underestimated the domestic water needs of rapidly growing western cities. Also, the vast wheat and corn belts of the mid-west have greatly depended not on reservoirs in distant mountains but on groundwater supplies, especially from the Ogallala aquifer. The gradual depletion of "fossil" water, a finite supply, in this and other western aquifers will lead to future crises in the agriculture of those regions.

The one-armed major pointed out that the tragedy of dam break floods will never stop the building of new dams for the benefits they provide.

> ... it is only the thoughtless man, governed by the impulse of the hour, and dragged from the throne of his reason by the emotions which arise at human suffering, who will believe that the vast industries which have been mentioned must be stopped because hydraulic power, when improperly controlled, may become an agent of destruction. Badly-constructed houses may fall and overwhelm families, but no check to the construction of houses will be made thereby. Fires will cause conflagration, yet homes will be warmed. Bridges may give way and trains leap into the abyss, yet bridges will be erected. Cars will leave the track and plunge travelers over embankments, but railroads will be operated. Dams will give way and waters overwhelm the people of the valley below, but dams will still be built.

Johnstown itself proves Powell correct in predicting the need for future dams, despite the 1889 flood. Sizeable dams exist in the watershed of the Little Conemaugh River, mostly to provide water supplies for communities in the region. And the

Quemahoning Dam, completed in 1913 about 20 km upstream from Johnstown, is one of the largest reservoirs in Pennsylvania. It covers nearly 880 acres and impounds 0.065 km³ of water in the Stonycreek watershed.

Powell's essay eventually returns to the South Fork dam. He believed the dam was properly built or repaired but assumed it had been improperly sited:

> Where, then, was the trouble? In the construction of the dam there was a total neglect to consider the first and fundamental problem—the duty the dam was required to perform. The works were not properly related to the natural conditions, and so a lake was made at Conemaugh which was for a long time a menace to the people below, and at last swept them to destruction.
>
> When the construction of such a dam is proposed, the first thing to be done is to determine the amount of water to be controlled and the rate at which it will be delivered to the reservoir under maximum conditions of rainfall or snow-melting.

Major Powell may not have realized the significance of all the changes made to the dam by the South Fork Fishing and Hunting Club, changes that were fatal to the dam. He would have known what was published in the engineering journals at that time. The failure of the dam, as modified and repaired by the Club, was only a matter of time. Powell will long be remembered not only for his contributions to exploration, geology, and the Civil War, but also for his research in the cultural anthropology of some western Indian tribes, where he served for a time as a Commissioner. In 1888 he was one of the scientists who helped establish the National Geographic Society. Most of his comments in "The Lesson of Conemaugh" resonate to this day.

Appendix 5: Newspaper Interview of James B. Francis while Attending the ASCE Annual Convention in Cresson, Pennsylvania

Johnstown Daily Tribune

Monday, June 30, 1890, page 1

THE SOUTH FORK DAM

Engineer Francis Says the Report Will Shortly be Made Public

The report of the committee that investigated the cause of the breaking of the South Fork dam will not be made public at this meeting of the American Civil Engineers' Society at Cresson. The examination was made as soon as railroad communication was established after the breaking of the dam, and was very thorough. It was concluded by Mr. James Francis, Max G. Becker, Alphonso [*sic*] Fteley, and W. C. Worthen, all prominent men in the profession. Their report has been carefully prepared, and was sealed and handed in to the Secretary in January last, where it is now at the call of the Society [Author note: other sources reveal that the sealed report was kept by Becker].

Mr. Francis is now located in Lowell, Mass., and is one of the oldest engineers in the United States. In speaking of the matter yesterday he said he wished the report would be made public, as this thing of holding it so long was getting stale. He expressed as his opinion, however, that it would not be given out until the pending suits regarding the disaster were settled. Whether the reports would be favorable or not to the South Fork Club Mr. Francis did not say, but he did not hesitate to say that he did not approve of holding back the report so long. It was held, he says, at the instance of Mr. Becker, who first introduced the idea that they should not permit it to influence pending litigation, and the other members of the Committee had deferred to him. Mr. Francis is now in favor of having the report made public, but does not care to argue against the wishes of Mr. Becker, and as no other member of the Society cares to push the matter, it will be let alone. Mr. Becker is the engineer of the Pittsburgh, Cincinnati & St. Louis Railway, and resides in Pittsburgh. The opinion prevails that it is on account of his business associates in that city that he takes his stand.

Appendix 6: Membership in the American Society of Civil Engineers

In 1889 the ASCE changed its method of admitting new members. Perhaps this change was related to the controversy that arose over voting for president in the closing months of 1888 when an alternative ballot supporting Thomas C. Clarke was proposed over the nomination of Max Becker. The June 15, 1889 issue of *Engineering News* (p 550) commented on this change:

> Until last year the Society had this very serious defect in its organization, that any five members who happened to have a grudge against an applicant could absolutely prevent his admission to the Society by secret black-balls. So long as this was so, the Society was not in a position to say that if any engineer did not belong to it, it was his own fault, and, consequently, it could not rationally demand that membership should be made one of the essential evidences of good standing in the profession. This is so no longer. An amendment passed last year, largely as a result of discussion in this Journal, put a stop to secret black-balls. Any engineer in good standing, who has had the requisite experience, can now belong to the Society if he will. All engineers should do so, joining in the junior grades, if they have not as yet the requisite qualifications for the senior grade, and all should use their influence to have membership in the Society a condition precedent to responsible public appointments at least, and so far as may be, to responsible private appointments also. .... Reform the Society from within, by all means; in some respects it needs it badly; but do not stand on the outside and growl.

Of course the elimination of blackballs was an important change for the ASCE. Any organization that rejected new members by allowing anonymous blackballs from a small percentage of current members will inevitably exclude qualified people because of their race, religion, sex, politics, birthplace, or even physical appearance. The ASCE is to be commended for rejecting in the nineteenth century the use of blackballs.

Modern professional societies nonetheless retain vestiges from the days when members were admitted or rejected by voting. Each year they select "fellows" in recognition of members who have made notable and lasting contributions in their field. "Fellows" include men and women, and this relict title hearkens back to when many older societies were "men only" organizations. Procedures vary among societies. Typically "fellows" can be nominated by any member but are vetted and selected or rejected by committees of other fellows. I personally believe the title "fellow" is truly antiquated and should be replaced by a title without gender. "Master" or "Emeritus" come to mind, but my proposed title would be "Luminary," which literally means a person of brilliant achievement who inspires and influences others.

Appendix 7: Johnstown Flood and ASCE Investigation of the South Fork Dam

– Timeline of Events –

May 30, 1889	At 7:44 a.m. river gage on Little Conemaugh River read "1.0 foot above low water"
May 31	At 7:44 a.m. river gage on Little Conemaugh read 14 ft. above low water; at 10.44 a.m. gage showed 20 ft.; final report at 12:14 p.m. said "Water higher than ever known; can't give exact measurement."
May 31	South Fork dam begins to overtop at ~11:30 a.m.
May 31	As the dam starts to overtop, Col. Unger sends John G. Parke Jr. to carry a warning message about the dam to telegraph from South Fork. Parke leaves South Fork by 12:00 noon to return to the dam
May 31	Around noon, PRR chief dispatcher Charles Culp hands Robert Pitcairn a telegram that Pitcairn knows is from Club president Unger. "… *I understood that morning before I started* that the people at Johnstown were warned out by Mr. Unger…" Pitcairn leaves Pittsburgh around 1 p.m., his special car attached to the #18 train.
May 31	Starting around noon and over the next few hours, observations are made that the rivers had crested or were beginning to fall slightly (i.e., PRR station master C. P. Dougherty at South Fork)
May 31	Dispatcher Culp testified he thought a 2nd telegram came in after 2:00 p.m., after Pitcairn left, "… to inform the people at Johnstown that there was a probability of the dam bursting…" This message was from Wilson, Superintendent at the Argyle Mines in South Fork, and did reach Pitcairn.

May 31	When Michael Trump (2nd in command to Pitcairn at PRR) reached Johnstown he was told by agent Deckert "…that he had a report the [South Fork] dam was liable to break at any minute…"
May 31	Breach of the South Fork dam at 2:50–2:55 p.m., leading to the Johnstown Flood. John Parke also reported the dam broke just before 3 p.m.; Andrew Carnegie and wife get the news while attending Paris EXPO
May 31	Flood wave strikes South Fork; railway station clock knocked off plumb at 3:08 p.m.
May 31	Flood wave reaches Johnstown ~4:07 p.m.
Jun 4	William Shinn leaves Pittsburgh to travel to ASCE meeting in New York
Jun 5	At the regular ASCE meeting in New York, a committee is appointed to investigate the dam failure; includes three prominent hydraulic engineers (Francis, Worthen, Fteley) and ASCE president Max Becker. Becker is chief engineer of a railway controlled by the PRR (via stock). William Shinn, one of ASCE's directors (and former managing partner with Carnegie), speaks at length about the dam
Jun 6	Coroner's jury in Cambria County declares that owners of the South Fork dam are responsible for the "…fearful loss of life and property…" from the breach of the dam
Jun 7	Coroner's jury in Greensburg declares cause of death of corpses carried downriver to Westmoreland Co. Verdict was "…death by violence due to the flood…" from the breach of the South Fork dam
Jun 13	PRR rail line repaired, restoring service between Altoona and Pittsburgh
Jun 14–20	During this week, ASCE committee visits the South Fork dam, interviews local citizens, takes survey measurements, and tours flood route
Jun 22	On this Saturday afternoon, Vice President Fteley gives address to ASCE convention in NJ, including preliminary report on South Fork dam
Jun 25	Letter from W. Boyer (Club's grounds superintendent) published in *Johnstown Daily Tribune*; describes the summer resort, including its "fine clubhouse" with 47 bedrooms and amenities. He wrote that "…all parties between the lake and Johnstown were notified three hours in advance of the flood that there was great danger of the dam breaking, and that it probably would, and warning them to flee. But they would not heed."
Jul 6	Pitcairn's friend, J. P. Wilson (Argyle Coal Co.) tells story published in *Johnstown Daily Tribune*; he twice sent Dan Siebert to

	check on South Fork dam and warnings were heeded in South Fork. Wilson claimed to have telegraphed: "The dam is breaking. Look out!"
Jul 15	In Pittsburgh, at direction of Pitcairn, PRR attorney John Hampton begins collecting testimonies from PRR employees and others who witnessed the flood; Pitcairn's own testimony is detailed and compelling, but in places distorts history
Jul 30	James B. Francis' daughter-in-law Caroline, wife of his son James, passes away at age 51 (b. June 19, 1838)
Jul 31, 1889	*Engineering News* announces in its "personal" section that Alphonse Fteley sailed for Europe this date to begin a month's vacation.
Aug 22	Engineer & eyewitness John G. Parke Jr. sends letter to ASCE committee
October	William Shinn, former managing partner with Andrew Carnegie, nominated to become ASCE President
January, 1890	Max Becker steps down as ASCE President; Shinn elected to that office
Jan 15, 1890	South Fork investigation report is completed and submitted to ASCE; it is sealed, not to be released
Jan 15	Although no longer president, Max Becker is given custody of signed and sealed ASCE report (normally ASCE's secretary would retain the report); this handling of report must have been approved by new president Shinn
Late May-Jun	Robert Pitcairn and his asst. Superintendent Michael Trump serve on organizing committee for the 1890 ASCE convention in Cresson; no visits to the nearby wrecked dam are organized
Jun 26 or 27	At Cresson convention, Max Becker tells a reporter that the South Fork report will not be released to avoid ASCE becoming involved in lawsuits
Jun 29	At Cresson meeting, James Francis says sitting on the report is getting stale and he wishes it would be released soon. The report was being kept sealed at the insistence of Max Becker. Alphonse Fteley attends this meeting, William Worthen does not.
January, 1891	William Shinn steps down, Octave Chanute becomes ASCE President. Chanute is retired, now makes his living as a railway consultant.
May 21–24	James Francis is the sole member of the South Fork investigation committee to attend the annual convention in Chattanooga, TN

May 22 (Friday)	Octave Chanute delivers presidential address at ASCE's annual convention in Chattanooga. He does not mention the South Fork report, which will be presented at this meeting
May 23, 1891	J. Francis presents in Chattanooga the long-awaited report on the South Fork dam, 2 years after the flood
Jun, 1891	Report on the South Fork dam is finally published in ASCE's transactions, vol. XXIV, two years after the Johnstown Flood. It was intentionally delayed and contains strong evidence of meddling from outside the 4-man committee, especially by Robert Pitcairn

References

Blodget L (1890) The floods of Pennsylvania, May 31 and June 1, 1889, p A143–A149, in Annual report of the Sec. of Internal Affairs of the Commonwealth of Pennsylvania for the year ending Nov 30, 1889, Part I, p 223

Brua SA (1978) Floods of July 19–20, 1977 in the Johnstown area, western PA. U.S. Geological Survey Open File Rept 78–963, 62 p

Coleman NM, Kaktins U, Wojno S (2016) Dam-breach hydrology of the Johnstown flood of 1889 – challenging the findings of the 1891 investigation report. Heliyon 2(6), 54 p, https://doi.org/10.1016/j.heliyon.2016.e00120, e00120

DCNR (2018) Link to vertical accuracy reports for LiDAR data in PA. PA Dept. of conservation and natural resources. http://www.dcnr.state.pa.us/topogeo/pamap/documents/index.htm [Accessed 5 Apr 2018]

Dolnick E (2002) Down the great unknown: John Wesley Powell's 1869 journey of discovery and tragedy through the grand canyon. Harper Perennial [ISBN 0-06-095586-4]

Fang X, Thompson DB, Cleveland TG, Pradhan P, Malla R (2008) Time of concentration estimated using watershed parameters determined by automated and manual methods. J Irrig Drain Eng 134:202–211. https://doi.org/10.1061/(ASCE)0733-9437(2008)134:2(202)

Folmar ND, Miller AC (2008) Development of an empirical lag time equation. J Irrig Drain Eng 134(4):501–506

Francis JB, Worthen WE, Becker MJ, Fteley A (1891) Report of the Committee on the cause of the failure of the South Fork dam. *ASCE Trans.* vol XXIV, pp 431–469

Haktanir T, Sezen N (1990) Suitability of 2-parameter gamma and 3-parameter beta distributions as synthetic unit hydrographs. Anatolia: J Hydrol Sci 35(2):167–184

NPS (National Park Service),(2012). Investigator's annual report, permit number JOFL-2011-SCI-0001

NRCS (National Resources Conservation Service) (2010) Time of concentration, Chap. 15 in: Part 630, Hydrology National Eng. Handbook, Appx 15A

NWIS (USGS National Water Information System) (2015) USGS 03041000 Little Conemaugh River at East Conemaugh, PA. Data online at: http://nwis.waterdata.usgs.gov/usa/nwis/uv/?cb_00065=on&cb_00060=on&format=gif_default&site_no=03041000&period=&begin_date=2014-10-14&end_date=2014-10-21. [Accessed 22 Apr 2018]

PASDA (Pennsylvania spatial data access), (2013a). Download sites for PAMAP LiDAR: http://www.pasda.psu.edu/uci/SearchResults.aspx?originator=%20&Keyword=pamap%20lidar&searchType=keyword&entry=PASDA&condition=AND&sessionID=5775552642015710143015 and http://www.pasda.psu.edu/uci/DataSummary.aspx?dataset=1247 [Accessed 22 Apr 2018]

PASDA (Pennsylvania Spatial Data Access), (2013b). Metadata summary for PAMAP program 3.2 ft digital elevation model of Pennsylvania (based on LiDAR data). Available at: http://www.pasda.psu.edu/uci/MetadataDisplay.aspx?entry=PASDA&file=PAMAP_DEM.xml&dataset=1247 and http://www.pasda.psu.edu/uci/DataSummary.aspx?dataset=1247 [Accessed 22 Apr 2018]

Powell JW (1875) The exploration of the Colorado River and its canyons. Dover Press, New York. 0-486-20094-9

Roland MA, Stuckey MH (2008) Regression equations for estimating flood flows at selected recurrence intervals for ungaged streams in Pennsylvania: U.S. Geological Survey Scientific Invest. Report 2008–5102, 57 p

Roussel MC, Thompson DB, Fang X, Cleveland TG, Garcia CA (2005) Time-parameter estimation for applicable Texas watersheds. Developed for Texas Dept of Transportation, research project number 0–4696

Russell T (1889) The Johnstown flood, monthly weather review, May 1889, in US signal service. Monthly Review, p:117–119

Townsend TF (1890) Pennsylvania State Weather Service Bulletins, p A34–A142, in Annual Report of the Sec. of Internal Affairs of the Commonwealth of Pennsylvania for the year ending Nov 30, 1889, Part I, 223 p

USGS (1976) John Wesley Powell – soldier, explorer, Scientist US Geological Survey Information [circular] 74–24. https://www.nps.gov/parkhistory/online_books/geology/publications/inf/74-24/index.htm [Accessed 22 Apr 2018]

Worster D (2001) A river running west – the life of John Wesley Powell. Oxford Univ. Press. 0-19-509991-5

Index

© Springer International Publishing AG, part of Springer Nature 2019
N. M. Coleman, *Johnstown's Flood of 1889*,
https://doi.org/10.1007/978-3-319-95216-1

Printed in the United States
By Bookmasters